EMERGING PATTERNS
OF INNOVATION

The Management of Innovation and Change Series

Edited by Michael L. Tushman and Andrew H. Van de Ven

Emerging Patterns of Innovations: Sources of Japan's Technological Edge

Fumio Kodama, with a Foreword by Lewis M. Branscomb

EMERGING PATTERNS OF INNOVATION

Sources of Japan's Technological Edge

FUMIO KODAMA
Foreword by Lewis M. Branscomb

Harvard Business School Press
BOSTON, MASSACHUSETTS

Copyright © 1991, 1995 by Fumio Kodama

This edition published by arrangement with Pinter Publishers

All rights reserved
Printed in the United States of America

99 98 97 96 95   5 4 3 2 1

Library of Congress Cataloging-in-Publication Data

Kodama, Fumio.
    Emerging patterns of innovation : sources of Japan's technological
edge / Fumio Kodama ; foreword by Lewis M. Branscomb.
        p.   cm. — (The management of innovation and change series)
    Includes index.
    ISBN 0-87584-437-5
    1. Technological innovations — Economic aspects — Japan.   2. Japan —
Manufactures — Technological innovations.   I. Title.   II. Series.
HC465.T4K627   1884
338'.064'0952—dc20   94-26384
                                                                                  CIP

The paper used in this publication meets the requirements of the American National Standard
for Permanence of Paper for Printed Library Materials Z39.49-1984.

Contents

List of Figures

List of Tables

Foreword

Lewis M. Branscomb

Just as the U.S. economy began to show signs of steady recovery, the worldwide recession of the early 1990s reached its depth in Japan. The stresses in Japanese industry, which has enjoyed double digit growth, have been about as severe as those in U.S. firms. Full employment commitments have been compromised, capital budgets have been slashed, and production has moved offshore. American firms making semiconductors, automobiles, and steel have gained market share against their foreign competition.

Japanese parity with American technology in many industries indicates that the sources of Japanese success no longer reside in second-mover advantages, which seek to penetrate established markets from a lower cost position with technology adapted from foreign sources. Interest-rate differences have largely disappeared. Both nations have strong research capacities. Both are relying on suppliers outside their borders. Both are manufacturing in foreign countries to gain full market access. While government policies are influential in both countries, it is no longer the Japanese government alone that pursues an aggressive industrial policy. Indeed, the Ministry on International Trade and Industry (MITI) has steadily reduced the level of its industrial intervention, while the U.S. government has largely abandoned its laissez-faire stance and now seeks a means to revitalize the civilian industrial base. Variations in corporate performance from firm to firm substantially exceed systemic differences between most industries in the two nations.

Are the market-share gains that American high-tech firms have made against Japanese competition the transient results of a weak dollar and out-of-phase recessions? Or do they reflect the fruits of U.S. industrial restructuring and, perhaps, the reaching of limits in the Japanese quest for ever higher productivity? It seems likely that with the maturing of the Japanese economic environment, competitive outcomes will increasingly reflect the fundamental strengths and weaknesses of the systems of industrial innovation in both countries. Given the degree of parity in the competitive situation, long-term outcomes are likely to rest on performance fundamentals rather than on trade policies or indirect government subsidies. Thus, understanding the changes that have occurred in the ways technologies are created and deployed in support of business strategies is critical to both corporate and public policies. What characterizes the technological dimensions of Japanese industrial performance? This is the issue Professor Kodama addresses.

The generation and employment of technology in commercial innovation is a fundamental performance factor for several reasons. First, with the steady increase in manufacturing productivity, there is every evidence that technological advantages are overtaking labor costs and capital investment as competitive factors. As Professor Kodama points out, research and development investments in Japanese manufacturing industry already exceed capital investment in equipment. Second, the management of technology follows no fixed rules, and each industry tends to evolve its own paradigms for dealing with technology. Third, because the roots of technology lie in science and most of science is generated outside of the corporate sector, public policies and national institutions may heavily influence the generation and use of technology in a firm.

This book is not in the tradition of most books written in English about Japanese management. Most of these, which are written by Westerners, attempt to interpret Japanese industrial practices in a Western frame of reference. As a result, they tend to emphasize the role of executive management. Professor Kodama is not interested in theory X or theory Z about management. His background is not business school but mechanical engineering, and his approach is fundamentally empirical. In this way, he avoids imposing his own prejudices on a complex set of issues.

Kodama explores the evolution of six dimensions of technological dynamics in Japanese firms. He calls these dimensions "techno-paradigm" shifts. They are posed as hypotheses and are refinements of concepts put forth in his previous book *Analyzing Japanese High Technologies: The Techno-Paradigm Shift* (London: Pinter Publishers, 1991). This book was also published in Japanese as *Haiteku-no-Paradaimu (The Paradigm of High Technology)*, by Chuo-Koron in 1991. The Japanese edition won the Sakuzo Yoshino Prize for the best book in social science in 1991. In writing *The Techno-Paradigm Shift*, Professor Kodama relied primarily on an empirical analysis of industrywide economic data. In this book, Kodama explores his hypotheses with three tools, each of which has its own strengths and weaknesses. Once again, he relies heavily on the empirical analysis of industrywide economic data to identify paradigm-shift. Kodama then allows the reader to test his interpretation against the empirical facts of specific cases. Here, he takes full advantage of the testimony of top technical executives whose views on technology strategy and innovation management have been analyzed in a monograph we coauthored, entitled *Japanese Innovation Strategy: Technical Support for Business Visions* (University Press of America, 1993). Finally, he reviews the evidence for the reflection of each hypothesis in both public and private policy.

These are powerful ideas, still in relatively early stages of evolution. Readers of Kodama's 1991 book will find that he has enriched his hypotheses and explored relationships among them. Kodama takes a systemic view of industrial technical activities, a quite different perspective from the Western, segmented, disciplinary view that is traditional in the United States. He puts manufacturing and evolutionary development at the center of the system and sees the conceptual integration of all the industrial functions as the key strength of Japanese firms.

Technology fusion—the evolution of technical paradigms drawing on multiple disciplines and integrating process and product technologies—is central to Kodama's paradigms. Manufacturing companies define themselves in terms of industry-specific or even firm-specific technology systems that become the basis for articulated business vision. Thus the idea of a manufacturing company as a knowledge-creating organization is not divorced from its practical function as a maker of things. Rather, success in manufacturing (and selling) is the practical expression of a unique, business-driven technological concept.

Kodama's treatment of diversification also emphasizes the importance of innovation strategies, for he sees them rooted in what I characterize as a "trickle-up" innovation process. Many Japanese firms introduce new technical ideas in mass-produced consumer products, first establishing the process technology through large-volume, low-cost production and after a period of "functional learning," raising the sophistication of product to reach higher-priced markets. Mastering process technology early in the innovation process broadens the technology base of the firm, reshapes the fusion of technologies, and provides a platform for product diversification if needed to protect the basic business strategy. This approach is in marked contrast to the Western approach of introducing new technology in the most demanding applications so that high profit margins pay for technology learning and small volumes decrease the stress on immature process technology.

According to Kodama's hypotheses, *demand articulation,* the systematic search among technological options for a means to satisfy attractive markets, drives the evolution of fused technologies. Demand articulation is a deeper idea than the simple notion that markets drive innovation. It suggests that a study of customer needs (as distinguished from market demand) leads to the identification of unrealized industrial capabilities that should be the basis for technology strategy. In this way, technology strategy can be bold and pioneering but disciplined to a market yet to be realized.

Kodama also explores the world of technology diffusion and its dependence on social and institutional adaptation. This line of thought is consistent with Henry Ergas, who finds the innovation process dependent on dynamic evolution of relationships between the innovator and all the suppliers, distributors, and customers. Thus innovation is seen as a process of socioeconomic change, not a matter of inventions exploited unilaterally by a firm.

While there is a trend in all countries now toward leaner, less vertically integrated firms that are increasingly dependent on alliances with specialized firms that can respond dynamically to changing technical requirements, most large Japanese companies are highly sophisticated in their ability to collaborate with other firms while competing fiercely with them. Kodama has extended the principles for operating in this cooperate-and-compete world to collaboration among nations in technological ventures. Unlike cost sharing as practiced in the European Center for Nuclear Research (CERN) and task sharing as represented in the U.S. space station, Kodama proposes *option sharing.* Option sharing assumes that each participating nation has some unique ideas about the best way to proceed in a collaborative venture but is willing to collaborate with others because of the total cost and complexity of the venture. Each nation then launches alternative explorations of prototype technologies after agreeing that whichever project proves to be the most meritorious will be accepted and the final installation will be built by an international team in the country with the accepted prototype.

This concept comes from the Japanese approach to orchestrated cooperation, in which robust competition is the central element. Such an approach can be observed in most of the major industrial research collaborations orchestrated by MITI. A three-year study of Japanese research consortia by Dr. Gerald Hane found that the Westerners do not understand how MITI operates. Most believe that MITI encourages precompetitive research collaboration in order to accelerate sharing of technical knowledge at the expense of early phase competition. It is no surprise that Americans who believe this conclude that the practice is incompatible with laissez-faire economics and verges on violations of anti-monopoly principles. However, as Hane has shown in *Issues in Science and Technology* (January 1994), MITI organizes the information flows and incentives in these collaborations to maximize effective competition among the key firms selected for the project. Kodama's option sharing is very much in that vein.

At the beginning of his book, Professor Kodama asks the most interesting question of all: Did Japanese executives invent the management tools and institutional innovations that comprise the emergent paradigms, or did changes in the technological innovation inherent in the evolution of technology provide the opportunity for historical social and institutional traditions to express themselves in particularly powerful ways? That is not a question of purely academic interest, for the dynamics of the processes that generate technology and govern its uses are ever-changing. If technological change intersected traditional Japanese institutional strengths in the period from 1960 to 1990, thus creating a unique opportunity for Japanese industrial success, that evolution can easily move beyond traditional Japanese industrial strengths. Does this really matter? It does indeed, for if the techno-paradigm shifts Kodama explores result from the accidental confluence of technology change and Japanese cultural/institutional characteristics, a great deal of adaptability may be required in the future.

Kodama's answer to his own question is a modest one. Japan benefited enormously from an evolution of the innovation process that matched the institutional capacities of Japanese society and the individual ethos and culture of the Japanese people. Perhaps, as new patterns of innovation emerge, they will not fit Japanese circumstances as well as they did in the past. Perhaps Japanese firms will have to demonstrate even greater adaptability to technological change in the future. Still, given the profoundly interesting characteristics of Japanese innovation described in this book, competitors must remain highly respectful of the adaptive capacity of Japanese innovators. Thus, there are profound lessons to be learned from the Japanese experience that go far beyond the specific strategies analyzed in this book.

January 12, 1994

Preface

Bilateral trade negotiations have left Japan and the United States ill at ease with each other. Even though Japanese business leaders are convinced there are rational explanations for some of the practices that Americans have termed exclusionary, they remain frustrated in their attempts to offer explanations that sound rational to Western ears.

In Japan, discussion tends toward the pragmatic: if a system works well overall, it must be a good system. This approach tends to overlook such questions as why things run well, what conditions are conducive to a system's well-being, what is good about a system, and what is wasteful. In the international business community, however, discussion does not take place without analysis. In pursuit of logical analysis, international businesses initiate dialogue. To the Japanese, however, this dialogue has the appearance of arguing for the sake of argument.

With this understanding of cultural differences in mind, I have undertaken to offer various logical explanations of the ways Japanese high-tech firms manage innovation and set their technology strategy. I hope these explanations will provide a starting point for dialogue between Japan and international businesses, from which each can benefit.

To give international businesses some logical insight into its achievements, Japan should conduct a thorough analysis of Japanese technology management practices from the viewpont of world history. At the same time, international businesses should attempt to understand the roots of their Japanese competitors' success and capitalize on their understandings in order to profit from competitive international trade. However trade frictions are resolved, it is the intrinsic productivity growth and innovation rate of businesses that will ultimately determine the economic health of nations and of the world.

I wrote my first book, *Analyzing Japanese High Technologies: The Techno-Paradigm Shift*, which was published by a British publisher in the spring of 1991, in an attempt to offer a rational explanation of some Japanese practices. When I began the book, I used a formal modeling analysis because I thought that it would constitute a universal language. However, I soon recognized that this would not be enough in a study of technology. To secure universality, I concluded, I needed to express myself in a foreign language, English, because it would force me to be more articulate in logical reasoning than my mother tongue. Although I had wanted the subtitle of this book to be its main title, I could not persuade my publisher.

After I had finished the book, I realized that Japanese business managers might not have a chance to read my arguments. To give them the opportunity, I translated my English book into Japanese and added two chapters to it that dealt with my recent work. It was published in the summer of 1991 by Chuo-Koron Co. with the title of *Haiteku-Gijyutu-no-Paradaimu* [*The Paradigm of High Technology*]. At the time, I was not sure that Japanese managers would accept my arguments. Much to my surprise, my book was awarded the 1991 Sakuzo Yoshino Prize, which is awarded to the best book of the year in the field of humanity and social sciences. I am the first engineer ever to be awarded this prize. While greatly honored, I came to feel that bilingual publication did not prove the universality of my arguments. Therefore, I decided to go to the United States to give courses based on my book at U.S. universities. Thus, I accepted an invitation by Harvard University to become a visiting professor at the Kennedy School of Government. The initiative for this invitation was taken by Professor Lewis Branscomb, and together, we taught two courses in fall and spring semesters. About one-third of the students in these classes were from the Business School. As few students dropped the courses, I gathered that American graduate students had little difficulty in understanding my arguments, even if they did not agree with me.

In order to learn the response of undergraduates to my ideas, I taught one course with Professor Stephen Kline at Stanford University in an interdisciplinary program entitled "Value, Technology, Science, and Society." I again enjoyed the students enrolled in the class and was pleased that few dropped the class. When I was about to leave the United States, Harvard Business School Press, at the urging of Professor George Lodge, who had read my first book, asked me to write a book that would include all I had learned during my two years in the United States as well as my previous writings. Immediately after I returned to Japan, I began writing, but it has taken me almost a year to rewrite, correct, and to add a substantial portion on management of high technologies.

Among the many people who have helped me as I have worked

on this book, I would like to thank Professors Branscomb and Lodge at Harvard University, Professors Kline and Daniel Okimoto at Stanford University, and Mr. James Raphael, Director of Research at the Asia/Pacific Research Center, who organized the implementation of my course at Stanford University. I would also like to thank Mr. Nicholas Philipson and Ms. Natalie Greenberg of Harvard Business School Press, who have corrected and commented on the drafts of each chapter. By responding to their questions, I think I have substantially improved and strengthened my arguments.

Finally, I would like to thank my wife, Minako, who came to the United States with me. My deepest gratitude goes to my three children, Miki, Eri, and Tomohisa, who were high-school students when we arrived at Harvard. Frequent changes in school, from Japan to the United States and from the East Coast to the West, gave them many opportunities for learning and some difficulties in adjusting to school life. Without their spiritual support, this book would have never come about. It is only appropriate, then, to dedicate this book to them.

EMERGING PATTERNS
OF INNOVATION

Introduction

When geographical shifts occur in the balance of world industrial strength, those countries whose capabilities are overtaken often attack the newcomers with accusations of conspiracy. Certainly this has been the case in the shift in industrial strength from West (the United States) to East (Japan). Critics, following the arguments of revisionists who claim that the Japanese system is an exception to world history, have frequently complained of a Japanese conspiracy.[1] In many cases, however, the conspiracy is no more than the spontaneous creation (innovation) of new institutions by the newcomers. In time, such innovations become the models for a new industrial order.

Since these innovations are outside the experience of the previous industrial leaders, they can indeed seem to make up a conspiracy. Those who have created the new institutions do not realize how new their system is to others and therefore cannot effectively refute the charges against them.

This phenomenon is paradoxical: unconsciously a less developed country that is trying to catch up with advanced countries resorts to policies that are more progressive and modern than those of the advanced countries. Only later does the country begin to appreciate why it chose such policies. At the turn of the century, this situation existed between England, France, and Germany. In an essay first published by *The New Review* in London in 1897, Paul Valèry, a French poet, described how England and France were stirred, almost scandalized, by German progress:

> It was around 1895 that England began to be no longer insensitive to the pressure of German power at the essential points of her economic life and her empire. . . . The English mind never hesitates to alter what seems bad, but it can for a long time resist changing what was once good and still satisfies it. . . . But in an era constantly in a state of technical transfor-

1

mation, and in which nothing escapes the drive to innovation, it is not enough to preserve what is.[2]

The ambivalence the British felt toward German progress brought about several political reactions. A similar situation seems to exist now between Japan and the nations of the West. In fact, if Japan and United States were substituted for Germany and England in the preceding quote, it would describe the current situation. Indeed, Super 301, the legislation introduced by the U.S. government in the 1988 trade bill that deals with disputes between individual countries on a broad range of trading practices deemed unfair, would seem to exemplify the quotation. Super 301 allows the U.S. government to identify a particular trading practice as a violation of U.S. law even when such a practice is not proscribed by the General Agreement on Tariffs and Trade (GATT). Under the terms of Super 301, the U.S. government may retaliate on a unilateral basis.[3]

Valéry clearly recognized that such a situation was not new but intrinsic to world history. Therefore, he tried to develop a general theory that might explain this historical paradox:

> Anyone cannot fail to look further into the phenomena of German expansion for a more general meaning. It is the moment for *ideas*, for comparisons, for a tentative formulation of theories. . . . So, in Germany's success I see, the success of a *method*.[4]

He argued that Germany's success was not due to individual technologies and institutions, but to the superiority of the method that integrated technologies and institutions. He went on to note the necessary conditions for this methodological superiority:

> Method calls for true mediocrity in the individual, or rather for greatness only in the most elementary talents, such as patience and the ability to give attention to everything, without preference or feeling. . . . But of all these virtues, the following are the most interesting: a well-made method greatly reduces the need to invent. It makes research cumulative.[5]

Valéry found that mediocrity made German expansion possible, and he predicted that the same phenomenon would some day appear between Japan and the West. In conclusion, he wrote:

> One is handsome, one is a genius only to others. Japan must think that Europe was made for her. And by virtue of a rational scheme already in operation in Germany, we should doubtless see the final triumph of all

the mediocrity on earth. . . . And what a curious result, if the results of that new order of things were in every way more perfect, more powerful, more pleasant than what we have today.[6]

Although Valéry was quite right in predicting that Japan would become one of the technological world superpowers and that its superiority would derive from "method," his analysis of "the final triumph of all the mediocrity on earth" must be doubted. He, too, seemed to doubt this analysis as he confessed, "But . . . I do not know. I am merely unraveling consequences."[7]

Further evidence of his uneasiness can be found in a footnote he added when he rewrote this essay in 1925. The original essay had noted, "The great philosophers are dead, and there is no place any longer for great speculative scientists."[8] By 1925, however, Valéry could not help but add the following comment: "This sentence should be struck out; men like Einstein, Planck, etc., make it inaccurate, that is to say, unjust."[9]

History has shown that people are confused and uncertain whenever basic changes are occurring because it takes some time for them to comprehend the essence of these changes. Thomas Kuhn, the philosopher of science, used the expression "paradigm change," to describe radical changes in basic scientific thinking.[10] By analogy with the concept of scientific paradigm, we can use the expression of "techno-paradigm shift" to explain radical changes in technological organization.

TECHNO-PARADIGM SHIFT

There is no doubt that Japan's socioeconomic system has led to its international competitiveness in high technology, but is this because the Japanese system differs from the Western system, or is it because a paradigm shift is occurring with the technology?

Put another way, has Japan deliberately established a unique socioeconomic system that is efficient in producing civilian high technologies, or has it succeeded because the Japanese system happens to fit the newly emerging techno-paradigm? Through an empirical analysis of the generation, innovation, and diffusion of Japanese high technologies, I believe that a paradigm shift in technology innovation, driven by the rapid evolution of science and engineering, is occurring and favors the Japanese system.

For some time, experts have been pointing out changes in the basic pattern of technology innovation.[11] With the emergence of new high-tech industries, major changes are occurring in both corporate and government policies. These changes merit the label "paradigm shift" because they are everywhere, and they are so profound that they may make conventional wisdom in business administration, economics, and international relations obsolete. And they are all taking place simultaneously, not in isolation from each other.

These are changes that affect not only the definition of manufacturing companies but also their business strategies. There have been changes in the research and development (R&D) activities that generate high technology and in product development processes that bring high technology to the marketplace. Underlying these changes are changes in the basic pattern of innovation. The effects of these changes go beyond the manufacturing environment into social institutions. The widespread use of information technology in society is possible only after adaptation of many social institutions to the potential of the new technology.

From my studies in the past ten years which have tried to codify the Japanese experiences with high technologies, I can distinguish six dimensions of paradigm shift: manufacturing, business diversification, R&D competition, product development, innovation pattern, and societal diffusion of technology.

Manufacturing

So great is the paradigm shift that is occurring in high technology that manufacturing companies, as the primary participants in technological innovation, have found the fundamental definition of the corporation is changing. Traditionally, a manufacturing company has been defined as a group of people who produce high-quality products at the lowest possible cost, using the most advanced equipment. Indeed, economists formulate manufacturing by means of the production function: capital plus labor make things. According to recent statistics, however, R&D expenses are greater than equipment investment in many Japanese high-tech companies.[12]

Japan embarked on the transformation of manufacturing after World War II. The process, which began in the 1950s and lasted through the mid-1970s, began with the importation of technology for making nylon, transistors, and televisions. During this time, a fairly high percentage of all Japanese R&D funds was spent on digesting the imported technology. From 1975 to 1985, Japan focused on developing technology to fuel the country's economic growth. As new technologies such as integrated circuits, liquid crystal display, and carbon fiber, were developed, capital investments were made in manufacturing to take advantage of them. This in turn led to economic growth, which allowed more R&D and kept the cycle going.

Since the mid-1980s, however, R&D investment has overtaken capital investment as an aggregate in the Japanese economy. This signals a dramatic metamorphosis of manufacturing, which may be characterized by a fundamental redefinition of the manufacturing company. When R&D investment begins to exceed capital investment, the corporation can be said to be shifting from a place for *production* to a place for *knowledge creation*. While the first two stages (importing technologies and developing technologies), show a linear progression, this ongoing change suggests that a paradigm shift is taking place, representing the third stage of Japan's management of technological innovation.

A new approach is needed to manage the knowledge-creating company. The view, deeply ingrained in Western management, of the organization as a machine for "information processing" will not suffice. A knowledge-creating company needs to be much more than that. It needs the ability to respond quickly to customers, create new markets, develop new products, and dominate emergent technologies. This necessitates a more holistic approach that views the company not as a machine, but as a living organism that can have a collective sense of identity and fundamental purpose.[13] The use of figurative language, once considered appropriate only for advertising campaigns, is now common in Japanese high-tech companies and may be an effective way to articulate and propagate tacit oganizational knowledge.

Business Diversification

A closer look at the recent metamorphosis of Japanese manufacturing reveals a reversal in two kinds of expenditures: capital investment has decreased, and R&D expenditure has grown. In 1985 the two curves created by these expenditures crossed. Further study has revealed that the rapid increase in R&D expenditure is driven by an increase in the breadth of R&D activity, i.e., toward more encompassing research, rather than in its depth, i.e., toward more fundamental research. In other words, the increase in R&D expenditure is related to the company's business diversification strategy.[14]

The conventional paradigm for diversification is that the development of a generic technology automatically brings diversification by applying the technology to various kinds of products. In this view, a technology is developed first for a technologically demanding, high-end product, then extended for the use of less technologically demanding, low-end products through the development of low-cost, quality-controlled mass-production processes. According to this view, diversification is based on the "spin-off" principle.

One of the most conspicuous elements of high-tech development in Japan, however, has been the codevelopment of product and process technologies. The development of a product is conducted in parallel with the development of its production technology. According to this view, without opportunities to accumulate production experience, high-tech development is not possible. This implies that business diversification in high-tech industries should follow a trajectory almost opposite to the trajectory based on the spinoff principle. Following this new trajectory, localized technical knowledge is developed and applied to less demanding, low-end markets first. The first Toray's carbon fiber, for example, was marketed for the shaft of a golf club. As the technology was mastered, the development trickled up into the high-end market, airplane's tail wings. Thus, diversification based on high technologies follows the "trickle up" process.[15]

The alternative to developing in-house technological capabilities is to acquire a new core technology through merger or acquisition. This option, however, is less available in Japan than elsewhere because of historical and cultural barriers to domestic mergers and acquisitions. Although diversification through acquisition is faster than diversification through internal development, internal diversification can better accommodate the trickle-up approach because the firm first masters the technology and learns its strengths and limitations before committing the larger amounts of capital needed for launching a new business.

The management of trickle-up diversification can be formulated around the idea of *core competencies*, defined as "collective learning in the organization."[16] To succeed in trickle-up diversification, management must ask, "Is there any consistent business strategy that can extend the firm's core competencies outside its conventional domain?" Because a successful business strategy must support (1) the firm's market, (2) its technology, (3) its products, and (4) its customer relationships, it is possible to develop four types of trickle-up diversification strategy.[17]

R&D Competition

Because Japanese manufacturers have found business diversification an essential strategy to guard against market saturation and technical surprise from competitors, they give high priority to managing corporate R&D activities. The conventional theory of R&D competition is based on the principle of *dominant design*, which was originally developed to explain the automotive industry.[18]

According to this theory, when a new technology arises there is considerable uncertainty over which of the possible variants will succeed. After a period of time and competition, however, one or a few of the variants come to dominate the others. This dominant design enforces standardization so that production or other complementary economies can be sought. Effective competition then takes place on the basis of cost and scale as well as of product performance.

Recent R&D competition among high-tech firms in the microelectronics industry, however, seems to be following a diverging pattern rather than the converging pattern implied by the dominant-design theory. In today's technology innovation race, manufacturers introduce new products every three years, before the learning process on the technology of the preceding innovation is complete. Competition may be characterized as technology *predation:* new devices drive their predecessors completely out of the market within six years of introduction. With this kind of technological advance, corporate decisions on investment are not made on the basis of the rate of return. They are made according to the principle of *surf-riding:* companies have no choice but to invest in successive waves of innovation or be left behind by competitors. Investment must continue just to stay in the market.[19]

This pattern of competition is likely to change further. While today, the competitor is usually another company within the same industrial sector, in the future the competitor may be a company in a different sector. In effect, a firm may not know from which corner the next competitor will appear. Thus, high-tech companies will have to monitor not only direct competitors in their own sector but also so-called *invisible enemies*, firms in other industries. In short, R&D competition in the high-tech industry should be framed as *interindustry* competition rather than interfirm competition in a given industry.

Therefore, both the corporate managers and the R&D managers of high-tech firms have to deal with R&D competition. We can discuss this management problem using a dichotomy of hierarchical versus horizontal coordination.[20] Although hierarchical coordination between planning and implementation is a major concern in conventional industries, horizontal coordination between R&D, technology development, and marketing is crucial in managing high-tech firms. Some organizational innovation may be needed to cope with the predatory pattern and to discover hidden enemies as early as possible. Some

Japanese high-tech firms are in the process of producing organizational innovations.

Product Development

In the high-tech era, the key issue of technology policy is not how to make possible unprecedented technological capabilities, but how to put technology to the best possible use. In the past, technology policy has emphasized the supply side of development; now it must work on the demand side.

Most descriptions of the process of technology development have employed the pipeline metaphor: new technology emerges from the successive steps of basic research, applied research, exploratory development, engineering, and manufacturing. Another, if extreme, view of technology development states that companies compete only on the basis of existing products, not those yet to be created. In this view, companies should seek constant, incremental improvements only in their products.

In reality, neither the pipeline view nor the picture of incremental development is adequate by itself.[21] Between these two extremes, however, is a wide range of product development processes in which some parts may be drawn from existing technical collections and some parts may be drawn from the pool of scientific knowledge. In fact, this range of processes may describe most of high-tech development.

In high-tech product development, therefore, the most important capability is the ability to convert demand from a vague set of distant wants into well-defined products, which I call *demand articulation*.[22] Articulating demand is a two-step process: first, market data must be translated into a product concept; and second, the concept must be decomposed into a set of development projects. Through the process of demand articulation, the need for a specific technology manifests itself, and R&D efforts are targeted at developing and perfecting that technology.

The concept of demand articulation becomes even more powerful when a national technology policy is analyzed. The development of the integrated circuit (IC), first in the U.S. defense sector and then in Japanese government-sponsored research consortia, best illustrates demand articulation at the national level. In the early development of IC technology, the U.S. government articulated and defined the problem to which the innovation should be addressed and supported promising aspirants in the development of that technology.[23]

Many companies in different industries were involved in bringing the integrated circuit from the defense sector into consumer-products market. In Japan, the government played a significant role in this transition by organizing a research association for very large scale integration (VLSI) development. When first formed, the association included all of Japan's major IC chip manufacturers, who then articulated their demand for manufacturing equipment and materials for chip making. In this way, an internationally competitive infrastructure was established.[24] This suggests that national policy can be discussed better using the concept of a "national system of demand articulation" rather than the oft-cited concept of a national system of innovation.[25]

Innovation Patterns

I have described the dimensions of the techno-paradigm shift from manufacturing through product development, from the level of the individual company to the level of national technology policy. The changing focus of manufacturing companies, the emerging method of diversification, the changing pattern of R&D competition, and the increasing importance of demand articulation are all related. But what lies behind these dimensions of the techno-paradigm shift?

For years it has been said that innovation is achieved by breaking through the boundaries of existing technology. Recent innovations in mechatronics and optoelectronics, however, make it more appropriate to view innovation as the *fusion* of different types of technology rather than as a series of technical breakthroughs. Fusion means more than a combination of different technologies; it invokes an arithmetic in which one plus one makes three.[26]

A number of revealing contrasts can be made between innovations achieved through fusion and those achieved through breakthroughs. First, while technical breakthroughs become possible when a prominent corporation in a specific industry takes a leadership role, fusion is made possible by joint operations among related industries. The mechatronics revolution in Japanese machine tools became possible through cooperation between Fanuc, a spin-off of a communications equipment manufacturer and developer of the controller, NSK, the bearing company that developed the perfect pitch ball screw, and a materials company that developed the teflon material used to coat the sliding bed. Second, while breakthrough innovations bring rapid growth for a particular corporation, fusion contributes to gradual growth in all the industries concerned. Third, while breakthroughs are often associated with defense policies, fusion is promoted through industrial policy.

The shift in innovation patterns—from breakthrough to fusion is implicit in the other dimensions of the techno-paradigm shift. There is a relationship between technology fusion and manufacturing companies' becoming knowledge-creating organizations. Technology fusion requires a synergy between different kinds of engineering subdisciplines. Such synergy induces substantial increases in R&D expenditure. Indeed, a company's R&D expenditures increase exponentially to the number of subdisciplines involved but its equipment investment increases only modestly by some linear function of subdisciplines.

Trickle-up diversification is a necessary condition for technology fusion. The management of trickle-up process should center on how to extend a firm's core competencies outside its conventional domain. Well-coordinated trickle-up diversification efforts among different companies in different industries could lead to technology fusion. If a firm seeks to attain business diversification through diversification of R&D, it can use the trickle-up process to establish a basis for technology fusion.

The shift in R&D toward interindustry competition will facilitate the realization of technology fusion. Because of the competitive threat from companies in other industrial sectors, some companies are forming alliances with companies outside their own sectors. Alliances between companies in different industrial sectors not only act as competitive hedges against technological surprises but also facilitate technology fusion.

Technology fusion is intrinsic to the process of demand articulation. R&D is demand driven, but companies do not always have the in-house technological capabilities to solve the technical challenges. To accumulate the necessary expertise requires a search and selection process outside the company. As companies develop their skill at articulating demand, they will also develop a skill at fusion.

Societal Diffusion

So far, my discussion has been confined to all aspects of the production of technologies and to a country's engineering infrastructure. The emergence of high technology, however, is changing the conventional wisdom about the interaction between technology development and society, i.e., how technology is best utilized not only by the industry but also throughout the society.

Freeman argues that the use of information technology throughout a society is possible only after many social institutions have undergone a period of change and adaptation.[27] Whereas technological change per se is often very rapid, there is usually a great deal of inertia in social institutions. The conventional model of societal diffusion is formulated around information-spreading mechanisms based on personal contacts. Although a technical adjustment process is sometimes built in the model, the diffusion pattern is viewed as epidemic in the sense of learning by infection; thus, its time-path follows the logistic curve.

The development of high technologies, however, is changing this pattern of diffusion and suggests that we need to look for signs of *institutional evolution* rather than for signs of a technical adjustment process. A shift in the societal diffusion process should be most conspicuous in the fields of telecommunication and welfare because these two fields are most affected by regulation. Japan offers at least two examples of the shift in the societal diffusion process. The diffusion of CAT (computerized axial tomography) scanners is a prime example of the relevancy of technology to social institutions accelerating rather than discouraging acceptance. Lobbying by a professional association of CAT experts brought about the government's decision to cover the use of scanners in its national health insurance plan.

A characteristic of information technology is its *network externality*.[28] Integration, however, requires some measure of technical standardization. In other words, the widespread diffusion of information technology requires the *coevolution* of technology and social institutions. Japan's unique database that tracks the installation of computers by forty-seven prefectural governments every year since 1963 reveals that the diffusion path is indeed epidemic, but to understand institutional coevolution, we need to study the forty-one categories of utilization, including payroll calculation, taxation, planning, and so forth, to discover any change in organization. Such a study will reveal that the diffusion of computer utilization is in fact a function of organizational change rather than technical sophistication. Furthermore, diffusion occurs more rapidly when the computers are used for new activities, indicating that institutional framework for new activities coevolves with the diffusion of the new technology.[29]

These findings may provide corporate managers with a logical base for implementing a variety of new technologies as they are developing structure and management processes. These findings may also help companies to respond positively to the technical challenges imposed by abrupt regulatory change. If regulatory changes are caused by societal needs, such as public concerns about environment, a company can implement technological change and still profit.

RESEARCH METHODOLOGY

Table I-1 summarizes the dimensions of techno-paradigm shift. Column 1 lists the specific dimension, column 2 describes the old paradigm, and column 3 gives the emerging paradigm. The remainder of the book will be organized around this table. Each chapter will correspond to one dimension of techno-paradigm shift.

Before beginning, however, I am going to describe the research methodology. Roughly speaking, there are two methodologies used in studying technological innovation: one is the case study, which is used by scholars in business school; the other is the formal modeling approach used by economists. Both methods seem to have intrinsic deficiencies.

A case study may generate insight and the insight can be surprising, but the findings are usually specific to the object and the environment of the case study. Policy implications can be drawn only through analogy. By comparing several cases, it may be possible to structure past experiences and thus learn important lessons, but if the whole paradigm is changing, past experiences will be of little use to decision makers, who are primarily interested in the future.

Moreover, case study data are collected through interviews with the people who developed the technology, but in many cases, engineers cannot relate the bounded rationality behind their decisions. They might have done something very innovative that they themselves are not aware of. This is especially true when paradigms are shifting. Usually, it is difficult for people to recognize their innovative accomplishments.

A formal model does allow us to generalize about past experiences,

TABLE I-1
Six dimensions of the techno-paradigm shift

Dimension	From	To
1. Manufacturing	Producing	Knowledge creation
2. Business diversification	Spin-off	Trickle-up process
3. R&D competition	Dominant design	Interindustry competition
4. Product development	Pipeline	Demand articulation
5. Innovation patterns	Breakthrough	Technology fusion
6. Societal diffusion	Technical evolution	Institutional coevolution

but technology is specific in every aspect even though science is universal. Therefore, there is always the risk of overgeneralization in a formal model. Technologists understandably complain about this approach, saying that it is difficult to identify what kinds of technologies model builders are talking about.

Furthermore, a formal model requires a rigid conceptual framework. Such a framework is often derived from past observable phenomena. Although established frameworks are the most appropriate for analyzing past technologies, when a new paradigm of technology emerges, there may be a serious discrepancy between the subject of study and the methodology used for the study. To study a new paradigm, we have to create a new conceptual framework. Indeed, Schumpeter made a breakthrough in the study of technology when he characterized the innovation process as a creative gale of destruction of the existing order, in which we might include the present frameworks for analysis of technology.

The individual wishing to study technological innovation may feel trapped between a rock and a hard place: case study without generalization is only storytelling, but a generalized approach without specific cases is not a study of technology. In what follows, I will try to strike a balance between these two conflicting approaches. This balance can be attained by taking a new approach to both case studies and formal models, but it would be effectively realized by bringing formal models closer to case studies. I will try to approach it from the policy science perspective, an area in which I have spent the past twenty years of my academic career.

The policy science approach is essentially a kind of engineering approach. Its aim is to obtain meaningful policy implications even if the approach is piecemeal. Using this approach, one develops the analytical tools most appropriate for a specific problem and then applies them to the available databases. In essence, this is a scientific endeavor associated with pragmatism: it deals with the dynamic aspects of the technology, but it departs from an academic obsession with a static, comprehensive framework that hopes for the elimination of all inconsistencies. In terms of technology policy analysis, this approach aims to acquire an understanding of the technology within an appropriate level of generality rather than an unobtainable, unified theory of technology.

Experience has shown that the policy science approach is easy to describe but difficult to implement. Scholars first advocated this approach almost twenty years ago, but few successful realizations have been produced.[30] As a result, scientific communities throughout the world doubt the possibility of implementing the approach. In order to regain the lost interest in policy science, we have to demonstrate the ideal type of analysis at which policy science had once been aimed.

Scholars are equally disillusioned with the management science approach. This approach stemmed from the success in operations research during World War II, but when it comes to management of technology, the approach has an intrinsic deficiency. First, managerial decisions about technology development are decisions at strategic levels, not operational decisions. These types of decisions must be made in highly unstructured environments.[31] Second, the process of developing a technology is not repeatable, because it is always a creative thought that leads to a business success. A business wins the competition only because it does not repeat a past approach or follow the strategies its competitors are pursuing. Third, there is no golden formula to follow for technology development. On an abstract level, the discounted-cash flow method may be applied to managing technology development, but in applying such a formula, it is often more crucial to estimate parameters rather than to make a precise calculation because the process of innovation is characterized by a high degree of uncertainty and unpredictability.

Last, but not the least, management science has failed to offer science policy administrators and research managers a vocabulary and a framework for talking about the choices they must make. Key concepts for these decision makers often are derived from an in-depth and sometimes naive study of technology rather than from an attempt to apply general management principles and mathematics to technology management. Therefore, we should be modest in talking about the management of technology. All that we can reasonably do is to provide an acceptable explanation about why and how some existing practices have proved effective. By doing so, we can at least put managers in better positions to invent new management approaches to the changing business environments in which high technology is going to play a pivotal role.

For the purpose of overcoming the difficulties that have confronted policy and management sciences, I have used case studies, mathematical formulations, and analyses of management practices throughout this book. Each chapter begins with specific case studies that are relevant to the chapter's subject. These will serve to create new vocabularies as well as to establish the basic mechanisms in each dimension of techno-paradigm shift. Then, I identify the most appropriate database for generalizing about the basic mechanisms discovered in case studies and describe a mathematical formulation of the problem. Although I have created most of the mathematical modeling, sometimes I have chosen the most appropriate model from among existing ones. At the end of each chapter, I introduce a few Japanese business practices that had been effective in managing each dimension of techno-paradigm shift and analyze why and how these management practices are accommodating the new vocabularies and concepts. In other words, using each dimension of techno-paradigm shift as a framework of appreciation, I will present what Nelson called an "appreciative theorizing" of successful business practices.[32]

By organizing each chapter in this way, I hope to join the three components of technology management analysis: case studies, formal modeling analysis, and appreciative theorizing. Although this oganization is still far from perfect, I believe that it will indicate future directions for the study of the management of technology.

NOTES

1.   C. Prestowitz, "Japanese vs. Western Economics: Why Each Side Is a Mystery to Each Other," *Technology Review* (May–June 1988): 27–36. J. Fallows, "Containing Japan," *The Atlantic Monthly* (May 1989):40–54.

2.   P. Valèry, "A Conquest by Method," in *The Collected Works of Paul Valèry*, trans. D. Folliot and J. Mathews (New York: Pantheon Books, 1962), 46.

3.   L. Tyson, "Managing Trade Conflict in High-Technology Industries" (paper presented at the National Academy of Engineering Symposium on Linking Trade and Technology Policies, National Academy of Sciences, Washington, D.C., 1991).

4.   Valèry, "A Conquest by Method," 51.

5.   Ibid., 58.

6.   Ibid., 65.

7.   Ibid., 66.

8.   Ibid., 59.

9.   Ibid.

10. T. S. Kuhn, *The Structure of Scientific Revolutions* (Chicago: University of Chicago Press, 1962).

11. C. Freeman, *Technology Policy and Economic Performance* (London: Pinter Publishers, 1987), 60–79.

12. F. Kodama, "The Corporate Archetype Is Shifting from a Producing to a Thinking Organization," *IEEE Spectrum* 27, no. 10 (1990): 82.

13. I. Nonaka, "The Knowledge-Creating Company," *Harvard Business Review* 69, no. 6 (November–December 1991): 96–104.

14. F. Kodama, "Technological Diversification of Japanese Industry," *Science* 299 (1986): 291–96.

15. L. Branscomb, "Policy for Science and Engineering in 1989: A Public Agenda for Economic Renewal," *Business in the Contemporary World* 2, no. 1 (1989).

16. C. Prahalad and G. Hamel, "The Core Competence of the Corporation," *Harvard Business Review* 68, no. 3 (May–June 1990): 79–90.

17. L. Branscomb and F. Kodama, *Japanese Innovation Strategy: Technical Support for Business Visions* (Lanham, Md.: University Press of America, 1993).

18. J. Utterback and W. Abernathy, "Dynamic Model of Product and Process Innovation," *Omega* 3, no. 6 (1975): 639–56.

19. F. Kodama and Y. Honda, "Research and Development Dynamics of High-Tech Industry," *The Journal for Science Policy and Research Management* 1, no. 1 (1986): 65–74. See also F. Kodama, "How Research Investment Decisions are Made in Japanese Industry," in *The Evaluation of Scientific Research*, ed. D. Evered and S. Harnett (New York: John Wiley & Sons, 1989), 201–14.

20. M. Aoki, "Toward an Economic Model of the Japanese Firm," *Journal of Economic Literature* 28 (1990): 1–27.

21. J. Alic et al., *Beyond Spinoff* (Boston: Harvard Business School Press, 1992), 19.

22. F. Kodama, "Technology Fusion and the New R&D," *Harvard Business Review* 70, no. 4 (July–August 1992): 72–75.

23. Organization for Economic Cooperation and Development, "Case Study of Electronics with Particular Reference to the Semiconductor Industry" (joint working paper of the Committee for Scientific and Technological Policy and the Industry Committee on Technology and the Structural Adaptation of Industry, Paris, November 1977), 133–63.

24. K. Oshima and F. Kodama, "Japanese Experiences in Collective Industrial Activity: An Analysis of Engineering Research Associations," in *Technical Coopera-*

tion and International Competitiveness, ed. H. Fusfeld and R. Nelson (New York: Rensselaer Polytechnic Institute, 1988), 93–103.

25. R. Nelson, ed., *National Innovation Systems: A Comparative Analysis* (New York: Oxford University Press, 1993).

26. Kodama, "Technology Fusion and the New R&D."

27. Freeman, *Technology Policy and Economic Performance,* 60–79.

28. P. David, "Technology Diffusion, Public Policy, and Industrial Competitiveness," in *The Positive Sum Strategy: Harnessing Technology for Economic Growth,* ed. R. Landau and N. Rosenberg (Washington, D.C.: National Academy Press, 1986), 373–91.

29. F. Kodama, "Can Empirical Quantitative Study Identify Changes in Techno-Economic Paradigm?" *Science, Technology and Industry Review* no. 7 (July 1990): 101–29.

30. E. Quade, *Analysis for Public Decisions* (New York: Elsevier Science Publishing, 1975).

31. As an illustrative example, see F. Kodama, "An Approach to the Analysis of Vocational Education and Training Requirements," *Management Science* 17, no. 4 (1970): B178–91.

32. R. Nelson and S. Winter, *An Evolutionary Theory of Economic Change* (Cambridge: Harvard University Press, Belknap Press, 1982), 46.

1

Manufacturing
From Producing to Knowledge Creation

Japanese manufacturing companies have experienced several major shifts in R&D investment since World War II. These shifts reflect distinct stages of industrial development, from import substitution to fulfillment of growing domestic demand to promotion of exports and avoidance of trade friction.

Japanese high-tech companies now spend on average 80 percent more on R&D than they spend on plant and equipment. A review of the recent pattern of investments for all Japanese manufacturing companies, however, indicates that this trend is not limited to high-tech companies. Capital investment has decreased while R&D expenditures have increased steadily. Considering this trend, we need to ask what the term *manufacturing company* means today. The largest Japanese firms seem to have entered a stage in which they survive by adapting to the changing environment through consistent, dependable R&D. Thus, the corporate archetype can be said to have moved from a goods-producing organization to a knowledge-creating organization that produces continuous innovation.[1]

An analysis of the postwar development of Japanese R&D activities will demonstrate how the transformation from goods-producing organizations to knowledge creating organizations was inevitable. A quantitative analysis will reveal three distinct periods: technology importation; technology development for economic growth; and today's transitional period toward knowledge creation.[2] An examination of the transition to knowledge creation reveals that the transition is more conspicuous in fabrication industries than in material industries.

But why have capital investments decreased while R&D expenditures have increased? One reason may be that soft automation (e.g., factory robots) has sharply reduced the need for capital expenditure. When a factory using soft automation switches over to making a new product, machinery is reprogrammed rather than replaced, and the cost of doing so is marginal. A flexible manufacturing system (FMS), then, has reduced industry's need for capital investment.

The perfection of FMS technology has freed manufacturing companies from certain burdens related to production and allowed them to increase R&D investment, which is related to future business. There are, however, unique problems associated with managing an organization in which knowledge creation is the major product. How does a company develop a collective sense of identity and fundamental purpose as it moves into new product lines? How do different units in a company communicate strategic decisions effectively? The final section of this chapter will discuss how several Japanese high-tech companies are solving these problems.

THE CHANGING INVESTMENT PATTERNS OF MANUFACTURING COMPANIES

In order to investigate changing investment patterns, we examined the capital investments and R&D expenditures of fifty companies. The companies were selected on the basis of the size of their R&D expenditures. Among the fifty-four companies that recorded the largest R&D expenditures in 1987, twenty-seven spent more on R&D investment than on capital investment.[3] To avoid the bias that might be caused by yearly fluctuations in annual capital investment, we can calculate each company's three-year average from 1985 through 1987 and compare it with the company's three-year average of R&D expenditure for the same period. The ratio of R&D expenditure to capital investment is shown in Table 1-1. For Hitachi and NEC, for example, these ratios are 2.04 and 1.30 respectively. In other words, Hitachi and NEC were spending 104 percent and 30 percent more on R&D than on plant and equipment.

The ranking of companies by R&D expenditure invariably includes such companies as NTT (Nippon Telegraph and Telephone Corporation) and Tokyo Electric Power. These companies primarily provide

TABLE 1-1
R&D expenditures compared with capital investment in major Japanese manufacturing companies (average, 1985–1987)

Company	R&D Expenditure (¥ million) (A)	Capital Investment (¥ million) (B)	A/B
1 Toyota Motor	258,333	301,333	0.86
2 Hitachi	251,773	123,367	2.04
3 NEC	235,667	180,667	1.30
4 Matsushita Electric	204,647	41,971	4.88
5 Toshiba	179,133	127,433	1.41
* NTT	178,987	1,674,388	0.11
6 Fujitsu	163,637	113,833	1.44
7 Nissan Motor	160,000	106,533	1.50
8 Honda	114,867	80,380	1.43
9 Mitsubishi Electric	113,000	77,000	1.47
10 Sony	92,978	77,968	1.19
11 Mitsubishi Heavy Industry	87,333	65,022	1.34
12 Mazda Motor	77,433	102,667	0.75
13 Nippondenso	59,860	86,233	0.69
14 Canon	59,563	53,579	1.11
15 Sharp	59,164	56,176	1.05
16 Nippon Steel	55,000	140,000	0.39
* Tokyo Electric Power	47,282	1,118,844	0.04
17 Fuji Photo Film	43,278	42,533	1.02
18 Sanyo Electric	41,945	56,261	0.75
19 Ricoh	36,720	26,671	1.38
20 Takeda Chemical Industries	34,900	22,440	1.56
21 Kobe Steel	34,233	71,865	0.48
22 Asahi Chemical Industry	33,833	52,033	0.65
23 Kawasaki Steel	31,267	100,133	0.31
24 Ishikawajima-Harima	31,267	12,933	2.42
25 Bridgestone	29,667	36,467	0.81
26 Sumitomo Metal Industries	28,671	79,133	0.36
27 Mitsubishi Kasei	28,000	31,200	0.90
28 Matsushita Communication	26,437	4,365	6.06
29 NKK	26,405	85,377	0.31
30 Fuji Heavy Industries	25,867	50,180	0.52
31 Matsushita Electric Works	25,200	24,480	1.03
32 Sumitomo Chemical	24,500	35,479	0.69
33 Oki Electric Industry	24,467	32,167	0.76
* Kansai Electric Power	24,139	587,000	0.04
34 Komatsu	23,233	14,433	1.61
35 Fuji Electric	22,698	18,038	1.26
36 Asahi Glass	21,000	45,667	0.46
37 Kubota	20,952	24,391	0.86
38 Kawasaki Heavy Industries	20,432	23,333	0.88
39 Sankyo	20,367	12,933	1.57
40 Victor Co. of Japan	20,326	29,187	1.01

TABLE 1-1 *(continued)*

Company	R&D Expenditure (¥ million) (A)	Capital Investment (¥ million) (B)	A/B
41 Sumitomo Electric Industries	20,167	27,833	0.72
42 Fujisawa Pharmaceutical	19,754	7,219	2.74
43 Eisai	19,382	5,927	3.27
44 Shionogi	19,191	8,786	2.18
45 Toray	18,100	41,850	0.43
46 Omron Tateishi Electronics	18,086	20,442	0.89
47 Konica	17,870	15,804	1.13
48 Aisin Seiki	17,145	23,389	0.73
49 Mitsui Toatsu Chemicals	16,295	22,536	0.72
50 Hino Motors	15,967	12,536	1.25
* Chubu Electric Power	15,562	188,331	0.03

Source: *Japanese Companies Handbook* (in Japanese) (Tokyo: Toyo Keizai Shimposhya, 1988).

Note: * indicates nonmanufacturing companies.

services, and their capital expenditures are much larger than their R&D expenditures. If these companies are excluded because they are not manufacturing companies, the average R&D/capital ratio of the remaining fifty companies is 1.29. In other words, on average, these companies spend 29 percent more on R&D than on plant and equipment.

Table 1-2 provides the total R&D expenditures and capital investments for all Japanese manufacturing companies from 1980 through 1987. The change in investment pattern remains observable. Furthermore, the time-series data indicate that this change has occurred only recently. According to column A/B of the table, the change occurred in 1986. While the ratio of R&D to capital investment was 0.62 in 1980, it became 1.17 in 1986 and reached 1.26 in 1987. The data in column B, however, includes capital investment for equipment and facilities related to R&D, such as construction of research laboratories and purchase of the research equipment. However, these investments are included in the R&D expenditures in the form of the depreciation costs. Therefore, investments on these items are counted twice.

To avoid such double counting, we turned to an annual survey conducted by the Japan Long-Term Credit Bank. The survey asks major companies to disaggregate their total amount of capital investment into categories by purposes: for an increase in production capacity, for labor-saving, and so forth, together with R&D-related items. Based on this survey, we calculated the average percentage of these R&D-related investments in capital investment.[4] This percentage is estimated for each year; for example, it was 6.0 percent in 1980 and 11.3 percent in 1986. Using these coefficients, we can estimate a capital investment that excludes investment related to R&D activities. This information is shown in column C.

A study of column A/C reveals that the reversal in investment patterns actually took place in 1985: ¥ 5.54 trillion for R&D expenditure and ¥ 5.47 trillion for capital investment. In 1986, these figures were ¥ 5.74 trillion and ¥ 4.34 trillion, respectively, and by 1987, they were ¥ 6.1 trillion and ¥ 4.15 trillion. The ratio of R&D to capital investment was 0.66 in 1980 and 0.90 in 1984. By 1985, however, the ratio had become 1.01. In 1986, the ratio was 1.32, and in 1987, the ratio reached as high as 1.47, meaning that Japanese manufacturing companies were spending 47 percent more on R&D than plant and equipment. Moreover, a closer look at Table 1-2 reveals that the change has been brought about by the opposite movements of these two types of investment: capital investment has decreased continuously since 1985 and R&D investment has grown since 1980. If capital investments and R&D expenditures were plotted on a graph, the two curves would be seen to have crossed in 1985.

Since we are interested in a paradigm shift of manufacturing companies, this phenomenon (R&D expenditure surpassing capital investment) should be analyzed in reference to individual companies rather than to the total of all the manufacturing companies. Therefore, we

TABLE 1-2
Time series of R&D expenditure compared wth capital investment for all Japanese manufacturing companies

Year	R&D Expenditure (¥ billion)	Capital Investment (¥ billion)		R&D/Capital Ratio	
	(A)	(B)	(C)	(A/B)	(A/C)
1980	2,896	4,651	4,372	0.62	0.66
1981	3,374	5,161	4,784	0.65	1.71
1982	3,756	5,099	4,727	0.74	0.79
1983	4,257	4,762	4,352	0.89	0.98
1984	4,777	5,788	5,285	0.83	0.90
1985	5,544	6,110	5,469	0.91	1.01
1986	5,740	4,896	4,343	1.17	1.32
1987	6,101	4,860	4,151	1.26	1.47

Source: Capital investment data from the Ministry of International Trade Industry; R&D expenditure data from the Prime Minister's Office of Japan, Statistical Bureau.

TABLE 1-3

Shift in cumulative distribution curves of R&D/capital investment ratios among the fifty largest manufacturing companies

Range of R&D/Capital Ratio	Percentile of Manufacturing Companies		
	1985	1986	1987
≤0.0	0	0	0
≤0.2	6	8	4
≤0.4	30	22	18
≤0.6	50	32	26
≤0.8	58	48	44
≤1.0	70	62	56
≤1.2	88	72	66
≤1.4	92	74	72
≤1.6	94	84	80
<1.8	94	86	82
≤2.0	94	92	86
≤5.0	100	100	100

Source: Japan Companies Handbook (in Japanese) (Tokyo: Toyo-Keizai-Shinpohshya, 1988).

   Note: Companies selected based on their total sales for each year.

are interested in an analysis of the distribution of R&D/capital investment ratios for individual manufacturing companies and in how general this phenomenon is in major Japanese manufacturing companies, not just R&D-intensive companies. It could be argued that choosing fifty companies on the basis of the size of their R&D expenditures biases the results in favor of R&D expense vis-à-vis capital investment. To eliminate bias, then, the criterion for choosing the fifty largest manufacturing companies became the total sales of a company. The time shift in cumulative distribution curves of the R&D/capital investment ratios among those companies is shown in Table 1-3 and Figure 1-1.

In this figure, if the curve shifts downward, R&D is exceeding capital investment in more companies. A trend in that direction is obvious. For 1987, the curve looks almost linear, meaning that the ratio is almost uniformly distributed among the fifty companies.

AN ANALYSIS OF JAPANESE POSTWAR DEVELOPMENT

How has the evolution of the Japanese manufacturing companies into knowledge-creating organizations happened? The results of a quantitative analysis of the R&D trajectory Japanese manufacturing has followed since the end of World War II will answer this question.

FIGURE 1-1

Cumulative distribution curves of R&D/capital investment ratios among the fifty largest manufacturing companies

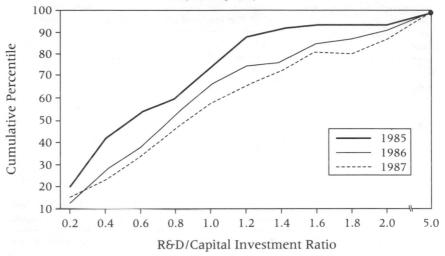

Our analysis reveals three distinct periods of development: *technology importation* from 1961 to 1975; technology development for *economic growth* from 1975 to 1985; and *transition* toward knowledge creation after 1985. In the first period, the major purpose of R&D was to digest the imported technologies.[5] In the second period, successful R&D resulted in capital investment, which led to economic growth.[6] In the period from 1985 on, R&D expenditure surpassed capital investment, resulting in a transition to knowledge creation.[7]

In order to analyze these structural changes, we need to choose an appropriate level and method of analysis for each period. The concept of unit of analysis is widely accepted by scholars in technology analysis.[8]

In the period of technology importation, digesting imported technologies was a national effort. Therefore, any analysis should be made at the national level. In the second period, the causal link can be adequately investigated at the individual R&D project level. The third period, the transition to knowledge-creating organizations, is concerned with the overall strategy of each manufacturing company. Therefore, the analysis needs to be conducted at the level of the individual firm.

Technology Importation: R&D for Digestion

In the decades that have followed World War II, Japan has moved from a country that imports technology to a technology oriented nation. The

TABLE 1-4

Percentage of R&D expenditure for technology digestion to the total R&D expenditure in each industrial sector

Food Manufacturing	1%
Textile	47
Lumber and Wood Products	29
Pulp and Paper	8
Chemicals	33
Chemical Fiber	72
Paint	5
Pharmaceuticals	7
Other Chemical Products	4
Petroleum Products and Coal	15
Rubber	69
Glass	15
Cement and Ceramics	7
Iron and Steel	18
Nonferrous Metals	16
Metal Products	15
Ordinary Machinery	17
Electric Machinery	21
Communication/Electronics	41
Automobile	15
Shipbuilding	4
Other Transportation Machinery	16
Precision Machinery	7
Other Manufacturing	1
Average	28

Source: Ministry of International Trade and Industry, Agency of Industrial Science and Technology (unpublished internal staff report, 1964).

seeds for today's R&D and technology lie in the technologies imported from abroad after World War II—nylon, transistors, televisions, and so forth. Unlike many other countries, however, Japan carefully screened, adopted, adapted the technologies it imported.

In 1964, MITI conducted a survey that showed the percentage of R&D expenditures used for digesting imported technology in each industrial sector (see Table 1-4). The survey covered 2,080 companies with R&D expenditures higher than ¥100 million. The sales of the companies surveyed composed 72 percent of the total manufacturing, mining, and utilities sectors. As shown in Table 1-4, as much as 72 percent of the R&D exenditure in chemical fiber was spent on digesting imported technology. In general, the material-related industries spent a larger share of R&D for digestion than other industries: 69 percent in the rubber industry and 47 percent in the textile industry. Fabrication industries, however, also spent a sizable amount of R&D for digestion: 21 percent in communication/electronics, 21 percent in electrical machinery, and 17 percent in ordinary machinery.

On average, the manufacturing firms spent 28 percent of their R&D expenditure for digesting imported technology in 1964. This percentage suggests that Japanese industry had a long-term plan for developing its R&D potential even in the early postwar period.

Government Controls

One of the most distinctive features of Japan's importation of technology was the government's extensive control. The Foreign Investment Act of 1950 required government approval for all transactions involving remittances in a foreign currency, and almost all technology imports involved such remittances. Applications were screened individually in some detail. Although the formal rationale for government controls was a concern for balance of payments, the controls affected the industrial structure and the composition of the technology imported as well as the price of technology.[9]

The controls gave priority to technology for intermediate goods, such as chemicals, and suppressed demand for technology for consumer-goods. The government's priority setting was most visible in MITI's 1950 list of thirty-three desired technologies, most of which were technologies for heavy and chemical industries. Only three technologies for consumer products were listed, and these were in pharmaceuticals, not consumer electronics.

Although each situation was different, the process of entry into an industry, importation of duplicate technology, and market growth can best be illustrated by a case involving the production of low-density, high-pressure process polyethylene, a raw material for the plastics used in such products as containers, furniture, and electric insulation.[10] In the 1950s, MITI singled out this particular product for domestic production.

The major goals of the government's policy were to avoid excessive competition among Japanese companies and to realize economies of scale, while at the same time establishing several competitors in each industry. These goals involved a subtle interplay of two objectives, one restrictive and one expansive. In order to achieve its goals, the government established a criterion for minimum economic scale for each company's production. At the end of 1964, the criterion for minimum scale of ethylene production was 300,000 tons annually.

Using this guideline, MITI approved licensing agreements one after another almost every two years.[11] In 1955, the first approval was given to Sumitomo Chemical, part of the well-established and large Sumitomo group, which purchased the technology from Imperial Chemical in the United Kingdom. In 1957, Mitsubishi Petrochemical, a new company created two years earlier by members of the large Mitsubishi group, which had an agreement with BASF in West Germany received approval. These two approvals were followed by others: Mitsui Polychemical (1960, DuPont), Nippon Unicar (1961, Union Carbide), Asahi Dow (1962, Dow Chemical), Ube Industries (1963, Dart Industries), Toyo Soda (1964, National Distillers), Nippon Petro-chemical (1965, Dart Industries), Showa Denko (1966, Ethylene Plastique), and Mitsubishi Chemical (1967, Gulf Oil). There are two discernible features in this entry pattern. First, half of these companies were created either as new companies or joint ventures for the importation of technology. Second, with the exception of Dart, Japanese companies chose a different foreign company each time as a source of technology.

Since the production capacity of polyethylene increased over thirtyfold from 1960 to 1970, the government's strategy had resulted in an increase in the number of producers but a decrease in the share of capacity held by the four firms (Sumitomo, Mitsubishi, Mitsui, and Nippon Unicar) that entered before 1962. By 1970, their share of capacity had decreased to 63 percent. Thus, we can say that the major goals of the government's policy (i.e., the subtle balancing of the two objectives) were successfully achieved.

However, the industry became so fragmented that the market share in 1970 of the four companies entering prior to 1962 was 17.9 percent, 17.9 percent, 13.9 percent, and 12.1 percent, respectively, while the tenth company retained a 3.4 percent market share.[12] By 1992, government control had resulted in even more fragmentation. Table 1-5 compares the largest Japanese chemical companies with those in the United States and Germany in terms of total sales and number of employees. It could be argued that the government controls accelerated fragmentation. Although members of the large industrial groups such as Mitsubishi, Mitusi and Sumitomo created new companies, some of these companies entered the market when domestic demand surpassed that forecasted, thus increasing the number of producers and fragmenting the market.

In other industries, however, government policies have adapted to developments within the industries.[13] In the case of the automobile industry, for example, government policies on foreign currency quotas and controlled trade during the industry's infant phase of development

TABLE 1-5
International comparison of major chemical companies, 1992

Country	Total Sales ($million)	Number of Employees
United States:	(17,202)	(59,000)
DuPont	37,208	125,000
Dow Chemical	18,971	61,000
Monsanto	7,763	34,000
Union Carbide	4,872	15,000
Subtotal	68,814	235,000
Germany:	(27,176)	(155,555)
Hoechst	28,420	178,000
BASF	27,585	123,000
Bayer	25,524	164,000
Subtotal	81,529	465,000
Japan:	(4,287)	(7,000)
Asahi-Kasei	7,814	18,000
Mitsubishi-Kasei	5,829	10,000
Sumitomo Chemical	5,004	8,000
Showa Denko	4,078	5,000
Ube Industries	3,897	8,000
Mitsui-Toatsu Chemicals	3,384	5,000
Mitsubishi Petrochemical	3,260	4,000
Toyo Soda	2,767	5,000
Mitsui Petrochemical	2,549	4,000
Subtotal	38,582	67,000

Source: For data on Japanese companies, *Japanese Companies Handbook* (in Japanese) (Tokyo: Toyo-Keizai-Shinpohshya, 1993); for data on U.S. and German companies, *Nikkei Annual Foreign Corporation Reports* (in Japanese) (Tokyo: Nippon-Keizai-Shimbunshya, 1994).

Note: Country's averages are in parentheses.

suppressed potential domestic demand to the level of the supply capacity of the domestic industry, while high tariffs suppressed the price competitiveness of foreign industry to the level of domestic industry. As the industry grew, tax measures and consumer-credit schemes raised the purchasing power of consumers, thereby allowing them to buy the products of the new industry. At the same time, the industry borrowed capital in excess of its borrowing capacity from government and government-guaranteed banks in order to expand production and bring down unit costs. In the industry's mature phase, production efficiency was raised through the accelerated depreciation of specified new machinery investments, and tax incentives for exports were used to enlarge external markets when the point of domestic sales saturation was reached.

Of course, it is always difficult to prove that a particular industry would not or could not have grown and developed without a national industrial policy. It is widely believed, however, that the government was both the inspiration and force behind the movement to chemical and heavy industries that took place during the 1950s.[14] In the high-tech industries, such as consumer electronics, however, it is interesting to note that government intervention was less drastic, less conspicuous, and less regulatory. This difference in the government's style of intervention may be due to differences in the nature of the technology or to differences in the time the industries developed.

Departing from Importation

To discover how long the period of technology digestion lasted and to find structural changes in the country's dependence on technology importation, we made a correlation analysis between technology import payment and R&D investment.[15] If a country is highly dependent upon foreign technology and its principal R&D activity is to digest imported technologies, there should be a high correlation between R&D expenditure and the amount of technology import payment.

On this causal assumption, we can calculate moving *correlation coefficients* for data sets of seven-year periods.[16] The Bank of Japan has recorded the amount of payment for technology importation on a yearly basis since 1960. Although the average lead time of technology development is about five years, we can use pooled data for seven years to obtain statistically significant results.[17] Each data set starts from the first seven years of observation and moves upward chronologically. For example, a correlation analysis is applied for 1960–1966, 1961–1967, 1962–1968, and so on.

Since we are using pooled data sets for seven-year periods, it is important to adjust all the values by using appropriate deflators. The deflator for R&D expenditure is readily available from the Japanese government.[18] However, there is no comparable deflator for technology import payments. Import price indices are heavily biased toward large commodity imports such as petroleum. Since technology import contracts are associated with machinery and equipment importation, we can use the import price index for machinery and equipment as the most reliable index for technology import payments.

The results of our analysis are shown in the third column of Table 1-6 and depicted by the solid line in Figure 1-2. As shown in the figure, almost all R&D expenditures were closely correlated to technology import payment (moving correlation coefficient above 0.90) until 1973.

TABLE 1-6
Time series of moving correlation coefficient

Basis year	Time Period	Moving Correlation with Technology Import	Moving Correlation with Commodity Export
1963	1960–1966	0.98	—
1964	1961–1967	0.97	—
1965	1962–1968	0.98	—
1966	1963–1969	0.98	—
1967	1964–1970	0.99	—
1968	1965–1971	0.99	—
1969	1966–1972	0.99	—
1970	1967–1973	0.97	—
1971	1968–1974	0.95	—
1972	1969–1975	0.92	0.85
1973	1970–1976	0.92	0.85
1974	1971–1977	0.86	0.89
1975	1972–1978	0.88	0.88
1976	1973–1979	0.84	0.87
1977	1974–1980	0.78	0.91
1978	1975–1981	0.87	0.95
1979	1976–1982	0.82	0.95
1980	1977–1983	—	0.96
1981	1978–1984	—	0.97
1982	1979–1985	—	0.97
1983	1980–1986	—	0.96
1984	1981–1987	—	0.91
1985	1982–1988	—	0.88

This suggests that the main purpose of R&D activity was to digest the imported technology. However, in the period from 1971 to 1976 (base year 1974), the correlation begins to fall, indicating the use of R&D for purposes other than digesting imported technology.

This change in the correlation implies that R&D gradually decreased its dependence on imported technology. Digestion enabled Japan to utilize the imported technologies and move into the next stage of development: domestic use and exportation. The departure from the dependence upon imported technology and the subsequent structural shift in R&D activity are confirmed by the analysis of R&D expenditures and commodity exports shown in the fourth column of Table 1-6 and by the dotted line in Figure 1-2. As can be seen in the figure, the moving correlation progressively increased from about 0.85 during 1972 to above 0.90 in 1977 and peaked in 1981 and 1982.

FIGURE 1-2
Change in moving correlation coefficients

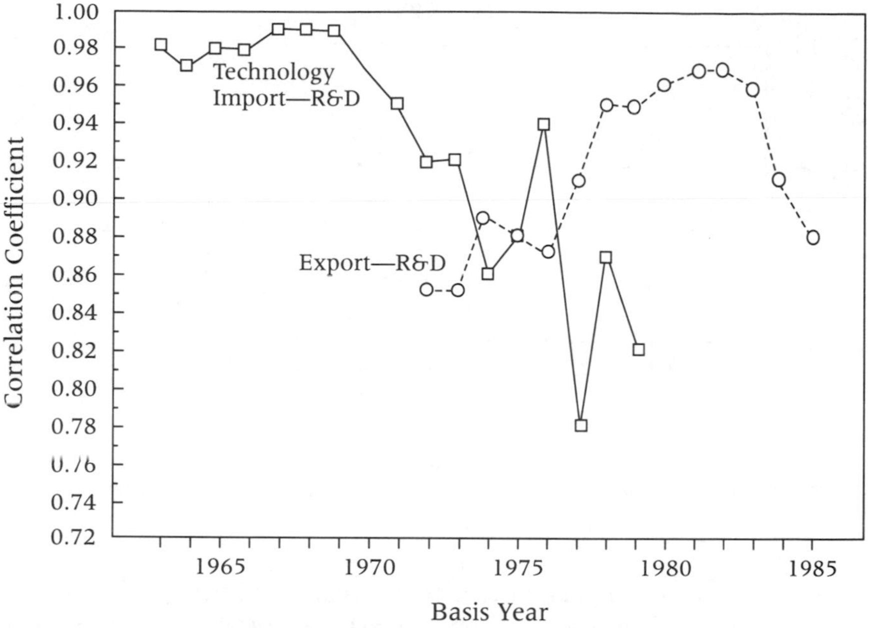

Basis Year

THE STRUCTURE OF THE METAMORPHOSIS

R&D for Economic Growth from 1975 to 1985

Once the imported technology was assimilated, R&D efforts could be directed toward economic growth. The dynamic relation between R&D as an input and economic growth as an output is assumed to be as follows: new technologies produced by R&D induce new requirements for capital investment; increased capacity brought about by capital investment leads to increases in production; increases in production provoke growth of the entire economy; high economic growth makes increases in R&D expenditure possible.

It would be too difficult to analyze this causal chain in its entirety. Therefore, we will confine our analysis to the hypothesis that successful R&D efforts lead to capital investment. To analyze this hypothesis, Kikuchi and Mori at the National Institute of Science and Technology Policy (NISTEP) selected twenty-eight projects from those that were conducted by Japanese companies with R&D investments larger than ¥1 billion in the period from 1975 to 1985.[19] The distribution of the

FIGURE 1-3

Relationship between R&D expenditure and capital investment for selected projects in 1975–1985

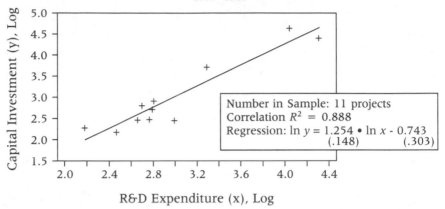

R&D Expenditure (x), Log

Number in Sample: 11 projects
Correlation R^2 = 0.888
Regression: ln y = 1.254 • ln x - 0.743
(.148)   (.303)

Source: J. Kikuchi et al., "Dynamics of Research and Development" (in Japanese), NISTEP Report No. 14 (Tokyo: National Institute of Science and Technology Policy, 1990).

period of R&D had a peak at four to six years, and the distribution of the time period between initiation of R&D and commercialization had a peak at five to six years.

In order to estimate the R&D expenditure for individual projects, Kikuchi and Mori used a survey to collect cost data in the form of the number of man-hours invested, materials/general expenses, and purchases of equipment for each project. Because the average duration of R&D activities is five to six years, price increases can be used to adjust annual costs.[20] By adding together all the cost items, they estimated the real total cost for each R&D project.

Their survey also asked companies to detail the amount of capital investment made as a result of the individual R&D investment for eleven successful projects of the twenty-eight projects. Data on capital investment were obtained. Since the survey asked only about the total sum of capital investment prompted by each project, they assumed that capital investment was done one year before the first sale of the product and used the deflator for capital investment to adjust the data. Once this is done, they can calculate the amount of capital investment in real terms.

Figure 1-3 illustrates the relationship between the real cost of an individual R&D project and cost of the capital investment prompted by the project. A regression analysis of R&D and capital investment yielded:

$$\ln y = 1.254 \cdot \ln x - 0.743, \qquad R^2 = 0.888,$$
$$\quad (.148) \qquad\qquad (.303)$$

where y = capital investment and x = R&D investment. At 1.25, the elasticity of R&D to capital investment was far above 1.00, indicating that R&D investment had an accelerating multiplier effect. Thus, R&D investment had a positive contribution to economic growth through increased capital investment.

In order to differentiate the period of 1975 to 1985 from the period preceding it, a similar analysis was made of the data collected in 1978 by the Japan Association of the Techno-Economic Society (JATES).[21] The JATES data included twenty successful projects carried out from 1961 to 1975. Their regression analysis yielded:

$$\ln y = 1.272 \cdot \ln x - 0.743, \qquad R^2 = 0.663$$
$$\quad (.214) \qquad\qquad (.568)$$

The fitness of this regression was much lower ($R^2 = 0.66$) than that of 1975–1985 ($R^2 = 0.89$), implying that the multiplier effect of R&D investment was less deterministic. In other words, success in R&D did not necessarily prompt hypereconomic expansion.

Although the analysis shows a difference in the structural relationships between the two periods, it cannot trace the transformation process. Therefore, we conducted another moving correlation analysis to test for a phase shift and used exports as indicators of economic contribution.[22] If there was indeed a transformation, export statistics should reflect R&D output.

This new analysis was based on the moving correlation between the amount of commodity export and R&D expenditure. Both values were converted to real values by the use of appropriate deflators. The result is shown in the fourth column of Table 1-6 and by the dotted line in Figure 1-2. The moving correlation coefficient progressively increased from about 1977, peaked over 1981 and 1982, and declined thereafter. This indicates that R&D effort from around 1977 until 1982 was directed toward product and process development of international competitiveness and hence toward economic growth. However, the results also show that around 1985 there was departure from this relationship. Although it is difficult to postulate using this kind of analysis whether or not the departure is permanent, we can assume there was a structural change in R&D effort after 1985.

R&D in the Transition Period Following 1985

Having attained a technology level comparable to other nations, Japan began to make a concerted effort to develop a lead in technology. To do this, it was essential to move from traditional areas of development into new areas of product and process development and organizational structure.

To analyze the structure lying behind the phenomena of R&D expenditure surpassing capital investment, we can use a classification scheme for industrial sectors. Since a detailed classification scheme for fifty companies might result in too few samples in each category for statistical inference, we adopted the following division: fabrication and nonfabrication industries. Because the majority of companies in the second category were material-related companies, this category was labeled "material industries."

Using this division, 60 percent of the fifty companies that are largest in terms of their sales belonged to the fabrication industries and 40 percent belonged to the material industries. The balanced distribution indicates that the classification makes sense by itself. On the basis of this classification, the change in industrial structure can be formulated as the product of two kinds of changes: one in the composition of these two categories in a country's industrial structure; the other in the change occurring within each category.

First, we investigated the change in Japan's industrial structure in terms of the share of each category in the national total for output of production. Each category's share in the subtotal of the fifty companies' total sales was calculated. In 1980, the share of the material industries was 39 percent, but it decreased to 26 percent in 1988. The share of the fabrication industries was 61 percent in 1980, but it increased to 74 percent in 1988. Therefore, we can say that the orientation of the Japanese industrial structure moved toward the fabrication industry during the 1980s.

Having established the demarcation between fabrication and material industries, we can see if there is any difference between the two groups in the shift toward knowledge-creating organizations. To do so, we can compare the two groups' cumulative distribution of R&D/capital investment ratios, as shown in Figure 1-4. As the figure clearly shows, the curve for the fabrication industries is almost linear, while the curve for the material industries is convex upward. This indicates that the phenomenon of R&D expenditure surpassing capital investment is not unusual among fabrication companies but is still rare

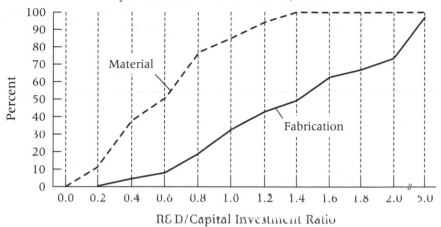

FIGURE 1-4

Comparison in cumulative distribution curves of R&D/capital investment ratios between fabrication and material industries, 1987

among material companies. The shift toward knowledge-creating organizations, then, has been realized only in fabrication industries.

The time shift in the cumulative distribution curve among fabrication companies during the 1980s is shown in Figure 1-5. As shown, the curve is shifting toward the bottom right corner as time passes, which implies that the phenomenon of R&D expenditures surpassing capital investment is becoming a general trend throughout fabrication companies. Note that the curves after 1986 are clearly different from those before 1985. This is additional evidence that the shift toward knowledge-creating organizations began about 1985. However, these changes are not as conspicuous in material companies as they are in fabrication companies. As shown in Figure 1-6, the curves stay at the top left corner of the graph. It is clear that the metamorphosis from production to knowledge creation is led by fabrication companies. The material companies seem to remain in the old paradigm, although some early changes can be observed.[23] Since the innovation pattern in the new materials revolution may be different from that in the conventional materials industry (to be discussed in chapter 5), a similar metamorphosis can be expected sooner or later in the material industry.[24]

Finally, we should ask if this reversal between R&D expenditure and capital investment will continue or will reach a saturation point soon. We asked the companies for their opinions.[25] The results are

FIGURE 1-5

Time change of cumulative distribution curves of R&D/capital investment ratios among fabrication companies

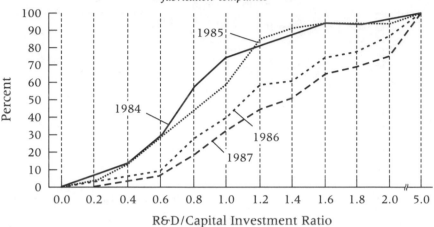

shown in Table 1-7. As seen in the table, 16 percent of the manufacturing companies queried responded "R&D will increase continuously," and 24 percent of them responded "increase above capital investment." Taken together, as many as 40 percent of the manufacturing companies surveyed believed the trend will continue. If the response category "parallel with capital investment" is included, then almost three-

FIGURE 1-6

Time change of cumulative distribution curves of R&D/capital investment ratios among material companies

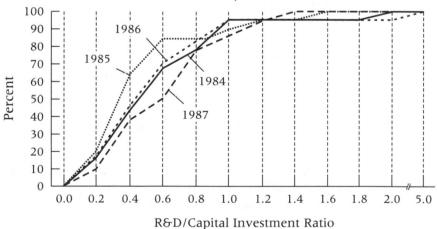

TABLE 1-7
The future trend in the relationship between R&D expenditure and capital investment

	Fabrication Industries	Material Industries	Total
R&D will increase continuously	16%	18%	16%
Increase above capital investment	28	18	24
Parallel with capital investment	36	24	31
Don't know	20	41	29

Source: Y. Kagita and F. Kodama, "From Producing to Thinking Organizations," NISTEP Report No. 15 (in Japanese) (Tokyo: National Institute of Science and Technology Policy, 1991).

quarters of the companies feel the trend will continue in the near future. At the very least they do not foresee a decrease in R&D expenditure, although it may reach saturation vis-à-vis capital investment.

As many as 40 percent of the material companies and 20 percent of the fabrication companies had reservations about this question, indicating that many companies are worried about a further substantial increase of R&D expenditure. These figures may provide us with good evidence that a paradigm-shift is occurring in the definitional aspect of the manufacturing.

FLEXIBLE MANUFACTURING SYSTEMS

Why is capital investment decreasing? Various forms of soft automation have provided manufacturing industries with the opportunity to reduce the amount of capital expenditure they need, even though both the variety of products and the volume of production are increasing.

An FMS is a computer-controlled group of semi-independent workstations that are linked by automated material-handling systems.[26] All activities—cutting metal, monitoring tool wear, moving parts from one machine to another, setting up, inspecting, adjusting tools, handling material, scheduling, and dispatching—are under computer control. The purpose of an FMS is to manufacture efficiently several kinds of parts at low to medium volumes.

The Emergence of FMS

While small-batch production of a diversified product range has been common in the clothing business for a long time, the term *FMS* became common only after it was applied to the metal-processing industry, where production is more complex.

In operation, an FMS is a miniature automated factory. System 24, initiated by Molins Ltd., a British firm, in 1967, is considered to be the first FMS in the world. In Molins' system, during the day, materials of various shapes would be set manually and held on pallets at a preparatory station. Once the materials were transferred to processing stations, unmanned automatic operation would be carried out day and night. At the time of its creation, the Molins system was not accepted economically.

Why then are flexible manufacturing systems a reality now? We would argue that both technical and market factors contributed to widespread FMS development. The technical factors included (1) the development of greater precision in mechanical technology; (2) the electronic revolution represented by microprocessors; (3) the emergence of robot technology; and, (4) the perfecting of quality control technology.

On the market side, FMS development seems to be inextricably connected with fierce intercompany competition and the saturated state of the markets for capital goods and consumer goods. A company cannot hope to survive in a saturated market with ferocious competition if it produces a standardized product or a product that is merely similar to those of its competitors. Such a strategy will only lead to either the maintenance of market share or the loss of market share without contributing to the overall growth of the market. For these reasons, manufacturing companies inevitably seek to give their products a competitive edge by a point of distinction.

An FMS permits a transition from the mass production of a small range of products to the small batch production of many different products. According to R. Jaikumar, U.S. managers are buying the hardware of flexible automation, but they are not using it correctly.[27] For the most part, management has treated the FMS as if it were just another set of machines for high-volume, standardized production. Furthermore, in the United States, an FMS cannot run untended for a whole shift, is not integrated with the rest of the factory, and is less reliable.

Jaikumar's survey, conducted in 1984 and based on more than half the systems installed at that time in the United States and Japan, yielded the following: the average number of parts made by an FMS in the United States was 10, while in Japan the average was 93; the annual volume per part in the United States was 1,727, in Japan only 258. In other words, U.S. companies used an FMS for high-volume production of a few parts, while Japanese companies used it for high-variety production of many parts at low cost per unit. It would also seem that U.S. FMS installations have not exploited opportunities to introduce new products. For every new part introduced into a U.S. system, twenty-two parts were introduced in Japan.

New Technical Dimensions

Although as a catch-phrase FMS seems simple and understandable, its technical nature is difficult to establish. Perhaps the easiest way to establish its technical nature is to compare an FMS with mass-production of a small product range. In Japan, manufacturers say *every 100 shots* fired in the technology of mass-production must hit the same bull's-eye, *every single shot* fired in FMS technology must hit a bull's-eye. Note the difference: in mass production, almost all shots must hit the *same* bull's-eye; in FMS production, each shot should hit the bull's-eye, but the *target* may change with each shot.

For a comparison of these two systems, let us assume an *ideal* FMS (each product is of different specification), and a single-product, mass-production system.[28] The comparison will deal with four aspects of each system: process accuracy; quality control; maintenance; and worker-skill requirements.

Process Accuracy. In a mass-production system, one can expect a *shakedown period* for the machines and a concurrent period familiarization for the workers as production experience increases. Once all the bugs have been ironed out of a single-product, mass-production system, one can also expect a certain latent production potential that is greater than what was expected. Latent production potential *cannot* be expected in an FMS.

From the viewpoint of process accuracy, the technological capability level of an FMS must be higher than that of the mass production by a factor of one or two at the start. An FMS can exist only when both machines and workers are capable of a higher level of process accuracy than they are in a mass-production system.[29]

Quality Control. The techniques of quality control (QC) were first developed in the United States. In the United States, the principle is to inspect every so many products at the end of the production line. This technique was revised in Japan to permit checks on the quality at each station of the production line and has proved to be a spectacular success, especially in the automobile industry. However, both techniques can only be used for quality control in a mass-production system.

In mass-production systems, QC procedures using either technique are followed after processing is finished. QC procedures for an FMS in which each individual product may be of different specification must occur based on signals received before processing. For example, a tool bit must be changed before it is worn out and causes product rejection. This is done by measuring the torque against preset standards.

Maintenance. An FMS offers its greatest benefits during night operations. Jaikumar found that eighteen of the sixty Japanese FMSs he studied ran untended during the night shift.[30] However, FMS maintenance requires a completely different concept of management than the concept required by a mass-production system.

For mass-production systems, it is important to minimize the downtime of the operation because they are daytime operations that involve workers. In the case of an FMS, downtime is preferable at night to a heap of junk products in the morning. When a substandard product is found, the entire system should stop and wait for maintenance workers to arrive in the morning. The key factor for FMS, therefore, lies in attaining such maintainability that the system stops the moment a substandard or reject product is found.

Skill Requirement. Workers at fully automated plants either use hoists to load workpieces onto the guideway to the machining center or prepare computer programs for the numerically controlled units. In other words, worker input is in the form of two separate kinds of labor—manual (the physical labor of using the hoist), and intellectual (the cerebral work of programming).

While many skills are involved in a mass-production system (welding, making electrical connections, or sequentially operating a number of machines, for example), all worker inputs fall within the definition of manual work. Thus, the concept of multiskill labor is different in a mass-production system and an FMS.

Knowledge Creation at the Factory Level

Because of the impressive achievements of an FMS in terms of cost and quality, engineers are, for the most part, absent from day-to-day operations of a plant. Rather than focusing their attention on internal operations, they focus their attention on the creation of fixed assets, such as systems and software. In his 1984 study, Jaikumar compared twenty-two FMS installations in Japan with the conventional computer numerically controlled (CNC) equipment they replaced.[31]

Jaikumar specifically compared the manpower requirements of various manufacturing systems for metal-cutting and noted that a conventional CNC-equipped factory needed 100 people but an FMS-equipped factory needed only 43. Even more interesting, however, was the change he observed in the composition of the work force. In the production line, an FMS cuts the number of workers from 60 to 22, while it cuts the number of staff from 40 to 21. Thus, an FMS makes the manpower requirements for a production line equivalent to that for the staff function. Within the staff function, moreover, the largest manpower reduction occurs in manufacturing overhead. Here, an FMS cuts the number of staff from 22 to 5, while in engineering, an FMS cuts the number only from 18 to 16. Thus in staff function, engineers outnumber production workers three to one.

With an FMS, then, the personnel composition of a company changes. Production is achieved with less manual labor, so there is a greater concentration in professional and intellectual management, and the key to competitiveness is not in producing more efficiently, but in maintaining a culture of innovation. This signals a fundamental shift from production to knowledge creation in the manufacturing environment, even at the factory level.

Does this mean that an FMS allows manufacturing companies to increase R&D investment? Jaikumar summarizes his arguments as follows:

> Each factory's profits will erode over time as other companies acquire the same operating capabilities. One way for a company to stay ahead is to create new physical assets in the form of better programmed and better managed equipment. Each plant's competitive fate would rest heavily on its ability to create facilities that generate performance advantages—and to do it faster than the competition.[32]

A substantial part of a company's ability to create such facilities is derived from its R&D efforts. Such tasks as the development of software and the prototyping and feasibility testing of manufacturing systems, may be conducted at a company's research laboratories and may be financed by a company's R&D budget. If this is the case, then a company's R&D expenditure increases while its capital investment decreases. An FMS allows a company to reduce its expenditure on capital equipment and focus resources on R&D and knowledge creation.

MANAGING THE KNOWLEDGE-CREATING COMPANY

What is a guiding principle for managing the manufacturing company when many of the burdens related to manufacturing are no more? Western management views the organization as a machine for "information processing." But the knowledge-creating company needs to be much more than that. It needs the ability to respond quickly to customers, to create new markets, to develop new products, and to dominate emergent technologies. This necessitates a more holistic approach, one that sees the company not as a machine, but as a living organism that can have a collective sense of identity and purpose.

According to Richard Nelson, to describe a firm adequately, one should focus on three strongly related features: strategy, structure, and core capabilities.[33] He notes that once the strategy is defined, one can expect the structure of the firm to be adapted to optimize the strategy and the core capabilities to be adjusted to both strategy and organization. R. Gomory, formerly the senior vice president for science and technology at IBM, associated the Japanese approach with "pulling technology" toward collectively defined goals, while U.S. firms push technology toward their goals.[34] The challenge for U.S. firms is, therefore, rapid product improvement in a manufacturing environment with severe constraints on both cost and quality. R. Frosch, vice president of General Motors, described the process of getting research into application as *navigation* between the two cultures of R&D and Operations.[35] The intrinsic conflict is that the culture necessary for R&D is diametrically opposed to the culture for operations. The focus on short-term payoffs, returns on investment, cost centers (business unit organization), and immediate profitability has created a culture in direct contrast to that required for technological innovation. In Japanese firms, the mechanisms for building bridges between these cultures are overt and built into a company's organization and its management practices. Nonaka asserts that cryptic slogans, which may sound more appropriate for an advertising campaign to a Western manager who thinks of management in terms of accountability and control, are in fact highly effective tools for managing the knowledge-creating company.[36]

Indeed, Japanese managers of high-tech firms often go to considerable lengths, including the adoption of slogans, symbols, cartoons, and metaphors for their businesses, to ensure that employees share a common vision of the firm, its mission, even its "creed and culture." They see the use of symbolism and figurative language as effective ways to articulate and spread tacit organizational knowledge.

The Use of Figurative Language

In the spring of 1992, a seminar, entitled "The Development of High Technologies in the U.S.A. and Japan: Comparative Evaluation and Policy Implications," was held at the Kennedy School of Government at Harvard University.[37] In their lectures, senior technical executives of several Japanese companies described with exceptional clarity the explicit relationships existing between business conception, corporate organization, and technology strategy in each company. Those in the audience were struck by how distinct these descriptions were. Using examples from the seminar, we will try to illustrate how Japanese firms are using figurative language.

NEC Corporation, the first international joint venture in Japan, was established in 1989 as a wholly owned subsidiary of AT&T's Western Electric Co. Its consolidated income was $26 billion in 1989. With 163,000 employees, 35,000 of whom are engineers, its major business areas are communications (25 percent of its total revenue), computers (44 percent), electronic devices (18 percent), and home electronics and other (13 percent).

NEC's logo is C&C (Computers and Communications), and it is more than a symbol for corporate identity. The C&C logo represents the entirety of NEC's corporate strategy, spanning both business and R&D strategies. For NEC, C&C represents the technological and market fusion of computers and communications. NEC uses the C&C logo for formulating and planning both its corporate strategy and its technology strategy.

The use of this logo as a central guiding principle helped NEC to identify chip making as basic to all its other developments and to deal with the diversity and chaos of subsequent R&D activities.

NEC's technology strategy—to obtain a leadership position in communications and computer technologies (C&C) to support the maximum number of markets that can sustain a focused technology strategy—is reflective of its self-image. As NEC has grown it has not diversified outside the scope of C&C, but it has extended the concept from industrial applications to business products to uses in the home.

In its attempt to emphasize technology as the root of organizational knowledge, NEC uses a tree to represent the company, as shown in Figure 1-7. The branches are the five product modules (communications equipment and systems, electronic devices, home electronics, computers and industrial systems, and "new opportunities"); the roots are the core generic technologies (materials, devices, systems, and software); and the sun represents the customer. It is an essential element of its strategy to keep the roots of the tree strong and healthy, even when the sun is not shining, i.e., when sales and profits are poor. In this metaphor, the upper part of the tree need only be healthy on average, but the roots must be kept healthy at all times.

Toshiba Corporation is a diversified firm that had $33 billion in

FIGURE 1-7
NEC tree

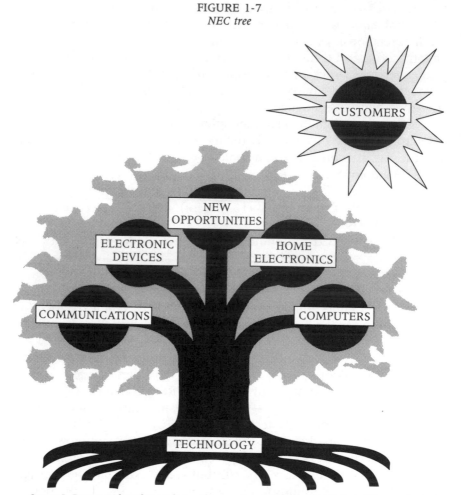

Source: L. Branscomb and F. Kodama, *Japanese Innovation Strategy: Technical Support for Business Visions* (Lanham, Md.: University Press of America, 1993), 51. Reprinted by permission.

consolidated business revenues in 1991. The corporation is horizontally integrated from nuclear power to consumer electronics; and vertically integrated from microelectronic chips to electric power systems. Its business covers digital electronics and components (57 percent of its 1990 nonconsolidated revenue), heavy electrical equipment for the utility industry (22 percent), and consumer products (21 percent). Forty percent of Toshiba's nonconsolidated revenue in 1990 came from chip making, making Toshiba one of the largest chip manufacturers.

Toshiba sums up its corporate identity as E&E (Energy and Electronics), viewing energy as the base of everyday life and electronics as a driving force behind today's highly sophisticated information society. Toshiba's history reflects the breadth of the company's technical vision. Its origins were the merger of the Tokyo Electric Co. with the Shibaura Engineering Works, which was itself the merger of an 1875 firm that made telegraph instruments (Tanaka Engineering Works) with an 1890 light bulb company (Hakunetsusha & Co.).

Toshiba operates four corporate R&D laboratories. The firm's emphasis on research is ascribed to Kisaburo Yamaguchi, who fostered basic research in the firm during the 1930s and came up with the adage "A manufacturer without R&D facilities is like an insect without antennae," i.e., research is needed to maintain a window on future trends and to cover future competition.

Toshiba describes its business strategy as the pursuit of CC&C (Competition, Collaboration, and Complementarity). This approach is most conspicuous in Toshiba's partnership with U.S. companies. The Toshiba-Motorola alliance, for example, is based on the exchange of Motorola's microprocessor technology for Toshiba's DRAM design and manufacturing technology. Toshiba's pursuit of CC&C is also evident in the agreement between IBM Japan and Toshiba to codevelop LCD display through a new joint venture company, Display Technology Inc. By bringing together complementary strengths, the companies involved in these partnerships fill out their product portfolios, enhance their overall technological capabilities, and compensate for in-house gaps and weaknesses.

Matsushita Electrical Industrial Co. Ltd., a large and diversified company, deals with audiovisual equipment, home appliances, and electronic equipment. Its products are marketed under the names Panasonic, Quasar, Technics, and National.

Kounosuke Matsushita, its founder, used a tap-water metaphor to describe the company's philosophy. In the terms of the tap-water metaphor, electrical appliances would be supplied to consumers with the same flow and abundance as tap water.

With this philosophy, Matsushita has always focused on consumer-related products, and Matsushita development engineers have been encouraged to work closely with their manufacturing, marketing, and sales counterparts. Matsushita has its own team for this effort known as *sei-kai-han-ittai* (united effort of manufacturing, development, and sales). Recently, teams of Matsushita engineers and sales employees visited 10 million households throughout Japan to listen to customers' comments about various consumer-related products.

In 1983, the company began to move into electric/electronic products and systems for businesses. Matsushita's medical imaging business is one example of its diversification into professional electronics. Although the R&D style, size, and research field of this business is outside of most Matsushita businesses, the company has continued to have engineering interacting with marketing, manufacturing, and sales. In this case, however, the relationship with the customer is different: the market is complex and hard to penetrate. Therefore, a Matsushita "developer-marketer" carries the product (or a prototype) to the customer site to receive one-on-one feedback.

Sony Corporation defines itself as the audio visual device supplier with the "finest sound and picture." Under license from Western Electric, its development of the first transistorized radio in 1955 took four years and an R&D investment of 12 percent of its annual sales. The development of the Trinitron TV picture tube took eight years, and the accumulated R&D expenditures were 19 percent of its annual TV sales.

Sony's greatest technical challenge was the development of a high-image quality electronic camera based on charge coupled device (CCD) technology. Begun in 1973, this thirteen-year project had a development budget that grew from 0.1 percent to 0.36 percent of total company sales between 1974 and 1983, a very high percentage for a single-product program. To encourage his engineers, Dr. Iwama, Sony's president at the time, identified a "big target" for their efforts. "Our key competitor is Kodak," Iwama said. With Kodak as the key competitor, the target was not just the replacement of imaging tubes in video cameras, but the creation of an electronic, rather than chemical, technology for making pictures.

Throughout Sony's history, corporate product vision has driven the development of specific technologies. At the same time, Sony has not attempted to cover all the technologies common to the consumer electronics industry. Considering Sony's self-definition as the supplier of audiovisual devices with the finest sound and picture, it is significant that it is not active in some technologies, such as thin-film transistors or liquid crystal, flat-panel displays, that might be considered necessary elements of a technology-driven consumer electronics company.

Sharp Corporation is a company that pays special attention to specific product goals that are important to the firm. It had consolidated revenues of $12.15 billion in 1992 and 65,200 employees, half of whom are overseas. Founded in 1912 by Tokuji Hayakawa, the company's first product was an "ever sharp" mechanical pencil. The company has become one of the world's leading consumer-product manufacturers, especially TV sets, calculators, solar cells, and TFT LCD flat panel displays.

R&D is decentralized to be near the product organizations. To integrate key technologies quickly into high priority new product lines and thus bypass the usual diffusion of technology from research laboratories to product development, Sharp has Gold Badge teams, which are established at the corporate level. Team members are chosen from corporate R&D, manufacturing group laboratories, and engineering. One member has the authority to run the team, and each member wears the same color badge as the company president, gold, signifying the priority the project has on company resources.

The team's work involves systems integration and technology exploitation. Each team is given a corporate priority to maximize speed to market, which is an essential competitive asset for a company like Sharp. A team typically stays in place for one and a half to two years, receives financial resources from headquarters, and has a priority call on personnel. On completion of the project, team members return to their original organizations. Although multidivisional task forces like these are well known, they rarely receive the level of visibility and priority accorded to Sharp's Gold Badge teams.

Sumitomo Electric Industries (SEI) began in 1897 as a family copper smelting business combined with silver extraction. By 1931, it was manufacturing copper wire and cable. The manufacture of cable for the transmission of information is the defining market for SEI. SEI's technology strategy is based on diversification through innovation to ensure that its position in a well-defined market is not threatened by unexpected innovations outside the industry. This strategy led SEI to recognize the emergence of optical fiber as both a threat and an opportunity to be acted upon. Today, it is the third largest optical fiber producer in the world.

To demonstrate a cumulative process of technological diversification based on internal development, SEI uses a bamboo garden metaphor (see Figure 1-8), which reflects the company's diversification objectives and growth strategy.

The bamboo garden suggests that SEI is going to grow many different stalks of bamboo (products) over the course of time. Bamboo propagates through underground roots, thus clumps of bamboo emerge separately from a common root system. The SEI bamboo roots are composed of both technology and market events, and the clumps of bamboo are integrated industry businesses, related to each other more by history than by mutual dependence.

As a guiding principle for diversification, SEI has a 50/50 program. The program maintains a 50:50 ratio between the company's main business and its diversified businesses. SEI is more diversified than

FIGURE 1-8
SEI bamboo innovation model

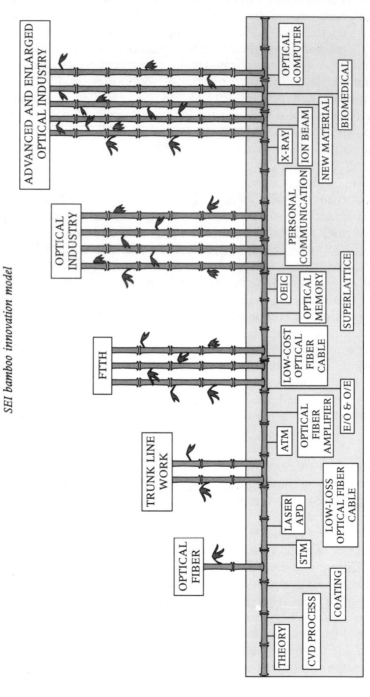

Source: L. Branscomb and F. Kodama, *Japanese Innovation Strategy: Technical Support for Business Visions* (Lanham, Md.: University Press of America, 1993), 52. Reprinted by permission.

it might be if it followed a technology strategy based on core technical competencies. Even though it says it is in the cable business, it has grown by branching into nuclear fuels, super hard metals, compound semiconductors, automotive brakes, and electronic systems. The common denominators for these fields are materials science and engineering. SEI always keeps cables in mind, but it does not try to minimize the number of technologies it produces. Nevertheless, electric wire and cable represented 40.7 percent of SEI's $15.87 billion in revenue in 1990.

Horizontal Coordination

Why do successful Japanese high-tech companies use logos and figurative language so widely, and how can their use lead these companies to rapid growth based on successful technology development while they have little effect on Japanese companies in other categories?

In his attempt to establish an economic model of the Japanese company and to provide a unified treatment of various features of Japanese practices, Masahiko Aoki compared the informational structures used by various Japanese companies in coordinating operational decisions among interrelated shops, with those used by Western companies.[38] He argues that an important internal characteristic of Japanese firms is horizontal coordination between operating units based on knowledge sharing rather than skill specialization. This is in contrast to the more familiar Western mechanism of *hierarchical coordination*, which is characterized by the separation of planning and implementation and by an emphasis on the economies of specialization.

In order to realize a horizontal coordination based on knowledge sharing, however, there must be some way to coordinate interrelated units. All the efforts of the Japanese high-tech companies just discussed can be interpreted as ways to make horizontal coordination possible. This type of horizontal coordination is not necessarily unique to high-tech companies. In fact, Aoki uses two specific examples from the automobile and steel industries. The highly publicized *kanban* system at the Toyota factory is used to facilitate a smooth flow between different workshops along the production stream, without the intervention of a supervisor. In the steel industry, an integrated engineering control room, which became necessary because of the switch from disjointed sequential to continuous steelmaking, is described as responsible for locating and solving cross-shop problems through discussion and bargaining with officials in charge of individual workshops.

The use of figurative language does not play a vital role in horizontal coordination in these two examples. In fact, in these two industries figurative language is used less frequently than it is in the high-tech companies. In the high-tech companies, however, the decisions were strategic decisions rather than operational decisions. Therefore, we can argue that the use of figurative language is an effective method for horizontal coordination in strategic decision making, while the development of an appropriate procedure, such as the kanban system, is vital in operational decisions.

CONCLUSION

Recent statistics indicate that R&D expenditures now exceed capital investment in many Japanese manufacturing companies, especially high-tech companies. These statistics give credence to the hypothesis that the fundamental definition of the manufacturing company is changing from a place for production to a place for knowledge creation.

A study of the statistics reveals that Japanese companies have undergone three distinct periods of development: technology importation; technology development for economic growth; and a transitional period to knowledge creation. During the last period, capital investment has decreased, as R&D expenditure has grown. Various forms of soft automation, such as the FMS, have reduced industry's need for capital expenditure sharply. Therefore, a paradox now exists: the perfection of sophisticated manufacturing technologies has reduced the burdens related to manufacturing, thus permitting an increased R&D investment, but this investment is related to future business. Some Japanese firms now link their survival to consistent, dependable R&D.

The technological capability level of an FMS should be higher than the technological capability level of mass production by a *factor of one or two*. Thus, the key to competitiveness lies in maintaining a culture of innovation, not in producing more efficiently.

The knowledge-creating company, however, requires new management tools to ensure that every employee shares a common vision of the firm. Hierarchical coordination, which separates planning and implementation operations and emphasizes economies of scale, may not be conducive to strategic decision making in a diversified firm. Japanese high-tech companies use figurative language to articulate and propagate tacit organizational knowledge. Figurative language can also be an effective method for horizontal coordination in the area of strategic decision making.

NOTES

1. I. Nonaka, "The Knowledge-Creating Company," *Harvard Business Review* 69, no. 9 (1991): 96–104.

2. F. Kodama, "R&D in the Past and Future Transition: Analyzing the Japanese Post-War Development," in *Economies in Transition: Statistical Measures Now and in the Future,* ed. P. Aven (Austria: International Institute for Applied Systems Analysis, 1991), 173–86.

3. F. Kodama, "How Research Investment Decisions Are Made in Japanese Industry," in *The Evaluation of Scientific Research,* ed. D. Evered and S. Harnett (New York: John Wiley, 1989), 201–14.

4. Y. Kagita and F. Kodama, "From Producing to Thinking Organizations," NISTEP Report No. 15 (Tokyo: National Institute of Science and Technology Policy, 1991).

5. Yada Mukdapitak, "Dynamics of Technology Importation: An Analysis of Japanese Experiences" (master's thesis, Saitama University, Japan, 1986).

6. J. Kikuchi et al., "Dynamics of Research and Development" (in Japanese), NISTEP Report No. 14 (Tokyo: National Institute of Science and Technology Policy, 1990).

7. F. Kodama, "R&D in the Past and Future Transition."

8. W. Abernathy and J. Utterbeck, "Patterns of Industrial Innovation," *Technology Review* (June/July, 1978): 58–64.

9. M. Peck and S. Tamura, "Technology," in *Asia's New Giant,* ed. H. Patrick and H. Rosovsky (Washington D.C.: The Brookings Institution, 1976), 535–58.

10. Ibid., 554–58.

11. Ibid., 555, Table 8-9.

12. Ibid., 557, Table 8-10.

13. F. Kodama, "A Framework of Retrospective Analysis of Industrial Policy," Institute for Policy Science Research Report No. 78-2 (Saitama University, 1978). A good summary of this report is found in Chalmers Johnson, *MITI and the Japanese Miracle: The Growth of Industrial Policy, 1925–1975* (Stanford, Calif.: Stanford University Press, 1982), 30–31.

14. C. Johnson, *MITI and the Japanese Miracle,* 31.

15. Y. Mukdapitak, "Dynamics of Technology Importation: An Analysis of Japanese Experiences."

16. These calculations can be found in Raman Letchumanan, "Technology and Trade Performance: Empirical Evidence from Japan's Experience" (master's thesis, Saitama University, 1991).

17. J. Kikuchi et al., "Dynamics of Research and Development," 11.

18. Prime Minister's Office of Japan, Statistical Bureau, *Report on the Survey of Research and Development* (in Japanese) (Tokyo: Nihon-Toukei-Kyoukai, 1970–1982).

19. J. Kikuchi et al., "Dynamics of Research and Development."

20. The methods used for these adjustments were as follows: (1) the times of starting and completion of R&D were identified respectively for each project, and the midpoint was calculated; (2) on the basis of the wage rate at this midpoint, the total man-hour cost was estimated by multiplying the amount of man-hours by wage rate; (3) by the use of a GDP deflator at the midpoint, the real cost for materials/general expense was calculated; and (4) by use of the capital investment deflator, the real cost for equipment was calculated.

21. Japan Association of Techno-Economic Society, "Evaluation of the Efficiency of R&D Investment (IV)" (in Japanese) (Tokyo: Japan Association of Techno-Economic Society, 1978).

22. R. Letchumanan, "Technology and Trade Performance," 87.

23. C. Freeman, "Technical Innovation in the World Chemical Industry and Changes of Techno-Economic Paradigm," in *New Explorations in the Economics of Technological Change*, ed. C. Freeman and L. Soete (London: Pinter, 1990), 74–91.

24. F. Kodama, "The Corporate Archetype Is Shifting from a Producing to a Thinking Organization" *IEEE Spectrum* 27, no. 10 (1990): 82.

25. Y. Kagita and F. Kodama, "From Producing to Thinking Organizations."

26. R. Jaikumar, "Post-Industrial Manufacturing," *Harvard Business Review* 64, no. 6 (1986): 69–76.

27. Ibid.

28. F. Kodama, "Preconditions for Flexible Manufacturing System: Technology, Industry, and Culture," *Technova Mechatronics News* 1, no. 1 (Tokyo: Technova Inc., 1983), 17–22.

29. H. Yoshikawa, *Robot and Human* (in Japanese) (Tokyo: NHK Books, 1985).

30. R. Jaikumar, "Post-Industrial Manufacturing," 72.

31. Ibid.

32. Ibid., 74.

33. Richard Nelson, "Why Do Firms Differ, and How Does It Matter?," CCC Working Paper No. 91-7 (Berkeley, Calif.: University of California, 1991).

34. R. Gomory, "From the 'Ladder of Science' to the Product Development Cycle," *Harvard Business Review* 67, no. (1989): 99–105.

35. L. Branscomb and F. Kodama, *Japanese Innovation Strategies: Technical Support for Business Visions* (Lanham, Md.: University Press of America, 1993), 69–71.

36. I. Nonaka, "The Knowledge-Creating Company," 96–104.

37. L. Branscomb and F. Kodama, *Japanese Innovation Strategies*.

38. Masahiko Aoki, "Toward an Economic Model of the Japanese Firm," *Journal of Economic Literature* 28 (1990): 1–27.

2

Business Diversification
From Spin-Off to Trickle-Up Process

According to the spin-off theory of diversification, the development of a generic technology will automatically bring about business diversification. Therefore, a technology is first developed for a high-end purpose and is then extended to less technologically demanding low-end goods that can be produced at low cost once the mass-production process is perfected. One of the most conspicuous elements of Japanese high-tech development, however, has been the codevelopment of product and process technologies. The development of a product is in parallel with the development of its production technology. This implies that business diversification in high-tech industries follows a trajectory that is almost the opposite of the trajectory described by the spin-off principle.

In high-tech industries, a *localized* technical knowledge is developed and applied to less demanding low-end markets.[1] As the technology is mastered and production experiences accumulated, development is directed toward the high-end market. L. Branscomb of Harvard University characterizes the approach taken by Japanese consumer electronics firms to the commercialization of new technology as a "trickle-up" strategy: a new technology is introduced in a consumer item rather than a high-end or industrial product in order to gain manufacturing experience at low functional levels and at low cost. At the same time, studies are conducted on functions that are necessary for products in higher-value markets. Only after the process and the technology is mastered is the technology introduced into markets with higher margins and more specialized applications.[2]

This strategy may not be unique to the consumer electronics industry. In order to develop a model of the trickle-up pattern, we will examine companies in three industries: electronics, new materials, and biotechnology. These case studies will reveal how different trickle-up trajectories can lead to business growth. At the same time, however, our studies will show that diversification through the trickle-up process is unique to the high-tech industries and is becoming their dominant mode of business diversification.

Our quantitative analysis of diversification will uncover the dynamics of the trickle-up process. The data for this analysis were supplied by the Statistical Bureau in the Management and Coordination Agency of Japan, which monitors corporate R&D trends. Every year the bureau asks companies to disaggregate their R&D investment into thirty-two product categories. Using this information, we can measure the degree of diversification of R&D investment in various product categories and make an industry-by-industry comparison. The negative correlation between sales growth of primary products and degree of diversification will show that Japanese companies are using diversification as a corporate survival strategy.

To measure the direction of technological diversification—either *upstream* (diversification into products that are inputs) or *downstream* (into products that are outputs)—we will combine the database from the Statistical Bureau with an input/output table. The results will indicate that trickle-up diversification is downstream, i.e., a technology developed by a component supplier is widely applied to products that are outputs. An analysis of twenty-one industrial sectors, which covers the entire Japanese economy, will prove that trickle-up diversification is unique to high-tech industries.

In his discussion of the product cycle, Raymond Vernon noted that U.S. firms were able to stay ahead of their imitators because by the time a foreign firm had mastered a technology and begun to apply it better or at lower cost, U.S. innovators had abandoned it for new technologies.[3] In Japan, however, mature industries are diversifying upstream, thereby offsetting decreased exports of finished products with increased exports of manufacturing machinery and equipment. It can be argued that the emergence of high-tech development makes this strategy viable for mature industries.

The management of trickle-up diversification can be formulated around the notion of *core competencies,* defined as "collective learning in the organization." In the management of trickle-up diversification, a company must design a consistent business strategy to extend its core competencies outside its conventional domain, but how can this be done? Every business strategy must support the company's market, technology, products, and customer relationships. Because a company may give preferential weight to one of these variables in its business environment, we will examine four types of trickle-up diversification strategy.

DIVERSIFICATION TRAJECTORIES

One of the most interesting questions to arise as we note how trickle-up diversification strategy has spread from Japanese electronics firms to other high-tech areas, including new materials and biotechnology, involves the source of the strategy. Does the strategy come about as a result of the nature of the technology, the nature of the market, or the nature of related scientific findings?

A review of the development of liquid crystal display in the electronics industry will show that trickle-up strategy comes from the nature of the technology, i.e., from the process of perfecting earlier scientific findings. A review of carbon-fiber development in the new materials industry will show that trickle-up strategy grew from the market development of high technologies, i.e., a new technology was pulled for an entry into the market in order to gain manufacturing experiences. In biotechnology, trickle-up strategy is the result of a localized discovery in microbial fermentation that has raised the amino acid industry to its present standing, i.e., the trickle-up strategy is related to the unique development pattern of bioscience.[4]

Technology Development: Liquid Crystal Display

Although Europeans discovered liquid crystals almost a century ago, the basic idea of using them in display devices came about only when RCA invented the dynamic scattering mode (DSM) in 1967. RCA demonstrated various prototype products, including a device displaying numerals and letters, a window curtain, still-picture display equipment, and a display panel for airplane pilots. All the products, however, were premature given the then-available technologies, and RCA gave up on its commercialization efforts.

At the time RCA was trying to commercialize liquid crystals, the standard for display technology was the *cathode-ray tube (CRT)*. A flat panel display was nothing more than a dream, and other technolog-

ical alternatives to liquid crystals existed, including electrolumines-
cence and plasma display. Manufacturers agonized over which to use.
Since RCA developed liquid crystal technology as a display method for
general purposes, it chose to stick with CRT technology, as did most
manufacturers of CRT screens.

There were, however, clear commercial applications for liquid crystal displays, and the correct technological choice was made by more
specialized manufacturers with technology goals suitable to such applications. For example, the watch manufacturer Seiko wanted to develop
an electronic quartz watch and needed a technology that would extend
the life of its product, reduce energy consumption, and allow for a
smaller, thinner display. For that reason, Seiko began research into
displays using little energy and soon focused on LCDs.

At the same time, the Sharp Corporation was caught up in stiff
competition in the development of pocket calculators. The company
realized the key to success lay in creating smaller, thinner products
that used little energy. With the development of integrated circuit (IC)
technology, the parts involved in calculation had become much smaller
and consumed little energy. Therefore, it seemed possible to develop a
calculator that did not need new batteries after every ten hours of use.
By focusing on liquid crystal displays and complementary metal oxide
semiconductors (CMOS), Sharp was able to extend the life of its calculators and to develop a card-size calculator.

The success of these companies in perfecting LCD technology and
the manufacturing technology needed for mass production demonstrated to the world the effectiveness of liquid crystals and established
the basis for further liquid crystal technologies. As shown in Figure
2-1, Sharp gradually expanded its application of LCD technologies as
various innovations followed, bringing larger screens, greater precision,
quicker responses, color displays, and greater legibility. In fact, Sharp
has become one of the world's largest leading consumer-product manufacturers, holding a major position in TV sets, calculators, solar cells,
and thin-film transistor (TFT) LCD flat panel displays. Today, the company controls thirty-eight percent of the worldwide liquid crystal display market, which is valued at over $2 billion, and the market is
expected to more than triple by 1995.

Market Development: Carbon Fiber

The trickle-up diversification strategy is emerging even in the materials
industry as new technologies allow engineers to custom design materials by manipulating atoms and electrons.

FIGURE 2-1

Development and marketing history of Sharp LCD products

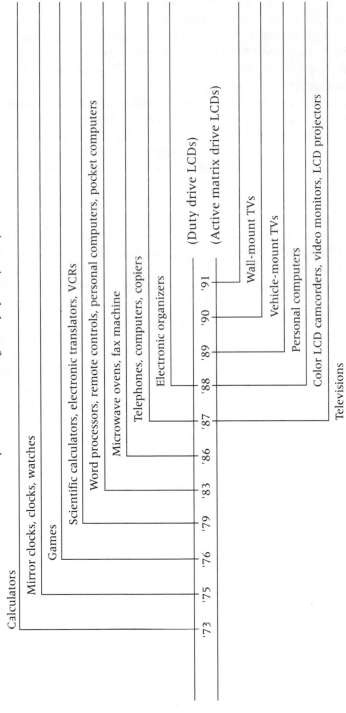

Source: Sharp Technical Library, 1 (1991): 66–67.

Many Japanese companies are already harnessing the power of this new generation of materials and trying to pull materials technology to the market in order to differentiate their products. Toray (one of Japan's leading chemical companies and a pioneer in new materials), for example, expanded its application of carbon fiber from golf clubs to airframes.

Union Carbide developed the first carbon fiber in 1959. In 1961, Dr. Shindo of the Japanese National Industrial Research Institute of Osaka found that polyacrylonitrile fiber could be carbonized under specific conditions. However, carbon fiber was not recognized as a new material until Great Britain's Royal Aircraft Establishment discovered that a high-strength, high-modulus carbon fiber can be created from acrylonitrile fiber because the graphite crystal generated in the carbonizing process is sintered under tension and arranged in the direction of the fiber. Rolls-Royce became interested in carbon fibers and tried to develop a jet-engine (RB207) that used composite materials reinforced with carbon fibers for its engine blades. However, the development of the engine was fraught with difficulties and led to financial crises for both Rolls-Royce and Lockheed Aircraft Co., the firm that was prepared to manufacture an aircraft with this engine.

Meanwhile, Toray became aware of the potential of the material and continued to collect technological and market information. In 1965, a researcher on organic synthesis at Toray's basic research laboratories made a vinyl compound, that was later found to promote carbonization of polyacrylonitrile fiber when it is copolymerized with acrylonitrile. Without these inventions, Toray would have never started the development of carbon fiber.[5]

In 1970, Toray invested ¥1.5 billion in a pilot plant. At the time, there was little demand for carbon fiber anywhere in the world. The market required barely one to two tons per month, while the pilot plant had a capacity of five tons per month. Toray, however, took the risk, recognizing that composite material reinforced with carbon fiber could be used in place of steel and light metals as a structural material. The properties of the carbon fiber obtained at the experimental stage were, however, uneven. After two years of developing its manufacturing technology, Toray's pilot plant produced carbon fiber with extremely little unevenness. The fiber was highly rated by the market, and Toray received many inquiries from customers around the world.

Although Toray decided to commercialize carbon fiber in 1971, it had difficulties in market development. There was no aircraft and aerospace market, the would-be primary market for the material, in Japan. In 1972, however, an American golf pro who used a club made

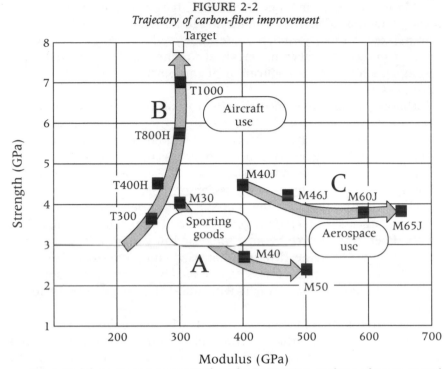

FIGURE 2-2
Trajectory of carbon-fiber improvement

Modulus (GPa)

Source: Yoshikazu Ito (paper presented at The Engineering Academy of Japan Second International Symposium on New Roles and Societal Status of Scientists and Engineers in the High Technology Era, Tokyo, 1990). Reprinted by permission.

of carbon fiber won a Japanese tournament. By the late 1970s, the demand for sporting goods made with carbon fiber began to grow rapidly.

The market for aircraft did not come about until the oil crises in the 1980s, when the improvement in fuel economy for civilian aircraft became imperative. Today, Toray is the largest manufacturer of carbon fiber in the world and has a net production capacity of about 2,500 tons per month. Toray's carbon fiber is the primary composite material used in fifteen percent of the structural materials in Airbus's A320 model. The material has significantly greater strength than comparable alloys and can be manufactured in one piece, eliminating the need for complex and costly assembly. Because of this manufacturing ability, the number of parts in Airbus's tail wing has dropped from 600 to 335. Toray is now developing carbon fiber for the automotive industries.

The difference in Toray's development trajectories can be most clearly seen in the strength-modulus diagram shown in Figure 2-2.

First, Toray developed carbon fiber for sporting goods by following path A in the diagram. Then, it developed the materials for aircraft and aerospace use by following the paths B and C.

Scientific Development: Amino Acids

Although biotechnology in the United States is targeted at producing pharmaceuticals, in Japan it has a wide range of applications. The Japanese take biochemistry for granted because of the country's long tradition of fermentation technology.

Monosodium glutamate (MSG) was discovered as the savory component of *kombu,* or seaweed, by Dr. K. Ikeda in 1908 and subsequently commercialized as a seasoning. During the early stage of the industry, MSG was extracted from gluten hydrolysate. Between 1932 and 1933, however, the increased market for MSG brought about an accumulation of an unsold by-product, wheat starch. To avoid this situation, a new extract process was developed, hydrolysis of defatted soybean. Some of the amino acids obtained from the hydrolyzed liquor were purified and marketed as reagents.

By the 1950s, there was a worldwide demand for MSG. To overcome the accompanying shortage of raw materials and the accumulation of unsold by-products, new processes were needed that could produce MSG at low cost and with a minimum amount of by-product. Vigorous research led to the development of chemical synthetic processes. The Ajinomoto Group pioneered a manufacturing method that used chemical synthesis.

In 1956, however, Dr. S. Kinoshita of the Kyowa Hakko Co. developed an industrial fermentation process of glutamic acid production. It was found that microbial fermentation was facilitated by the addition of cofactors. The principle of metabolic control evolved into a general technological procedure that could produce lysine or other amino acids by fermentation. The procedure not only brought about a major innovation in the sodium glutamate industry and in the fermentation field at large but also had an enormous impact on applied microbiology.

Rapid progress in research related to amino acid manufacturing methods using microbes followed. Processes ranging from a glutamic acid fermentation to a direct fermentation process, including a precursor process to avoid metabolic obstruction and an enzymatic process that is combined with chemical synthesis, came into existence. As a result, it became possible to produce almost all kinds of amino acids with microbes, as shown in Figure 2-3.

In recent years, technological progress, coupled with the biotech-

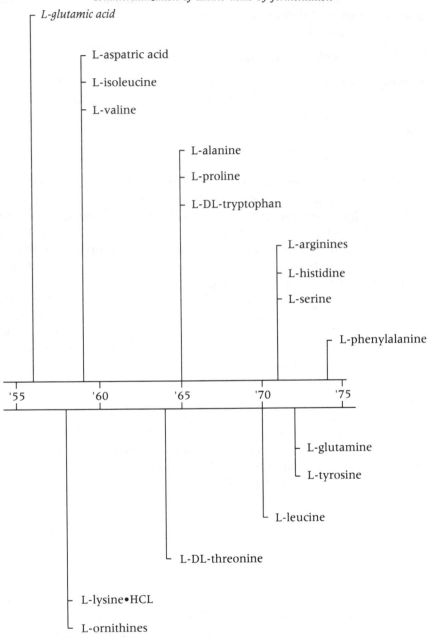

FIGURE 2-3
Commercialization of amino acids by fermentation

Source: Bioindustry (Japan), 7 (1990): 409–14.

nology boom, has spurred the adoption of fixation enzyme, cell fusion, and recombinant DNA processes. As amino acids are made available industrially, the manufacturing technology is accelerating even further.

From this example, it would seem that progress in basic biotechnology research is motivated by progress in industrial technology. Yet there are many cases in which progress in basic science has been driven by technological need. Therefore, we can argue that the trickle-up trajectory of biotechnology is embedded in the unique development pattern of bioscience.

A QUANTITATIVE ANALYSIS OF DIVERSIFICATION

There have been two previous attempts to identify a sectoral pattern of technological diversification. One in 1982 by F. Scherer, who studied U.S. patents, and one in 1983 by K. Pavitt, who studied significant British innovations.[6] The unusually rich R&D data found in Japan's *Report on the Survey of Research and Development*, which is published yearly, makes it possible for us to take a third approach.[7] The survey asks a company like Hitachi, for example, to break its R&D expenditures into such categories as chemical products, fabricated metal products, ordinary machines, household electric equipment, communication and electronic equipment, automobiles, precision instruments, and so forth.[8] There are in total thirty-two possible product fields. Because the survey is available for every year since 1970, a time-series analysis is possible and investigations at any particular point in time can be checked in terms of their particularity. Moreover, the survey is collected under the special law on statistics (*Shitei-Toukei*), so companies fill in the form they receive with great care.

R&D expenditure is often considered nothing more than one of the input indicators for business diversification.[9] However, unlike a sector's R&D expenditures within its principal product fields, R&D expenditure outside principal product fields can be assumed to reflect some of the results of business diversification. Because they are expenditures outside the core business activity, these investments can be canceled easily if the prospects are not favorable. Cancellation is not as easy in the case of investments within the principal product fields. Therefore, R&D expenditure in product fields outside a sector's principal product fields can be assumed to be output indicators for business diversification. (In chapter 3, we will elaborate the dynamic nature of R&D investment in more depth.)

TABLE 2-1
Classification of principal product fields for diversification studies

Industrial Sector	Principal Product Fields
Agriculture	agricultural; forest; fishing
Mining	mining
Construction	building construction and civil engineering
Food	food
Textile mill products	textiles
Pulp and paper products	pulp and paper
Printing and publishing	printing and publishing
Chemical products	chemical fertilizers and inorganic and organic chemical products; chemical fibers; oil and paints; other chemicals
Drugs and medicines	drugs and medicines
Petroleum and coal products	petroleum
Rubber products	rubber
Ceramics	ceramics
Iron and steel	iron and steel
Nonferrous metals and products	nonferrous metals
Fabricated metal	fabricated metals
Ordinary machinery	ordinary machinery
Electrical machinery	household electrical appliances; communication and electronics; other electric equipment
Motor vehicles	automobiles
Other transportation equipment	ships; aircraft; railroad; other transportation equipment
Precision equipment	precision instruments
Other manufacturing	other manufacturing products

Table 2-1 shows intramural expenditure on R&D disaggregated into thirty-two product fields for all Japanese companies with a capital of ¥100 million or more (3,803 companies in 1982).

Measurement of Diversification

Using the disaggregated data on R&D expenditure, we can draw a profile showing how each industry diversified during the 1970s. The textile industry is used as an example because it was one of the most diversified industries during this period.

The profile in Figure 2-4 shows product fields as a function of the industry's R&D expenses (on a logarithmic scale). Product fields are arranged in decreasing order of expense in 1970. The 1970 profile is a

FIGURE 2-4

Profile chart for diversification in the textile industry

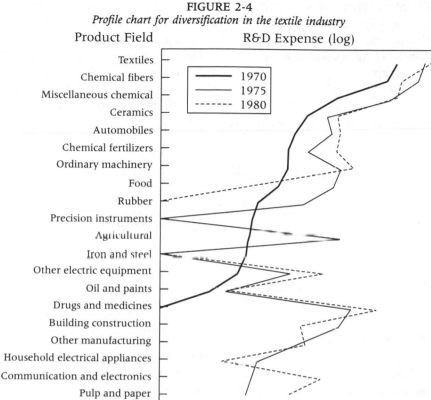

Source: F. Kodama, "Technological Diversification of Japanese Industry," *Science* 233 (July 18, 1986): 293. © AAAS.

monotone curve, whereas the profiles for 1975 and 1980 are no longer simple monotone curves, indicating that the rank order of expense in product fields has changed. The industry's R&D expenditure in building construction, for example, was nothing in 1970, was substantial in 1975, and had become one of the largest by 1980. There was no R&D investment in drugs and medicines in 1970, but it was the fourth largest expense item in 1980. To recognize the technological continuity that lies in this diversification, one need only to consider the textile firm of Asahi-Kasei, which now applies its fiber technology to building materials and to filters for medical equipment, such as those used in kidney dialysis.

Degree of Diversification

Using Table 2-1, it is possible to analyze quantitatively the degree of diversification in an industry. An industry's R&D distribution over different product fields (shown in the right-hand column of Table 2-1) can be supposed to be a probability distribution, as was visualized in the profile chart of Figure 2-4. Following this supposition, we can construct an indicator for degree of diversification, using the concept of entropy.

Given E_{ij} (*i-th* industry's R&D expense into *j-th* product field, for $i = 1, 2, \ldots, n; j = 1, 2, \ldots, m$) and allowing p_{ij} to be the share of *j-th* product field in the total R&D expense of *i-th* industry, *i.e.*,

$$p_{ij} = E_{ij}/(E_{i1} + E_{i2} + \ldots + E_{im}) = E_{ij} \Big/ \sum_j E_{ij},$$

then, H_i, the entropy of *i-th* industry, can be calculated by

$$H_i = - p_{i1} \cdot \log_2 p_{i1} - p_{i2} \cdot \log_2 p_{i2} - \ldots - p_{im} \cdot \log_2 p_{im}$$
$$= - \sum_j p_{ij} \cdot \log_2 p_{ij}.$$

If an industry's diversification is advanced and its R&D expenses cover many product fields with little concentration in specific fields, then the industry's entropy value, H_i, rises. If an industry's diversification is not advanced and its R&D expenses cover only a few product fields with high concentration in specific fields, then the industry's entropy value drops.

A time-series of entropy values of major industries (see Figure 2-5) shows that the textile industry is the most diversified, motor vehicle manufacturing is the least diversified, and that the iron and steel industry is in between. During the 1970s, the textile industry saw rapid diversification while the other industries did not.

There are several ways to interpret these results. It seems clear, however, that an industry's degree of diversification is related to its production growth rate in its principal product field (textile products in the textile industry and automobiles in motor vehicles manufacturing). The greater an industry's production of principal products, the lower its degree of diversification and less drastic its change (motor vehicles, for example). On the other hand, the more saturated an industry's production of principal products, the greater its degree of diversification and its change (textiles, for example). This suggests that diversification may be used as a survival strategy for a mature industry. This is especially true in Japanese industry, where the life employment

FIGURE 2-5
Measurement of diversification of major industries

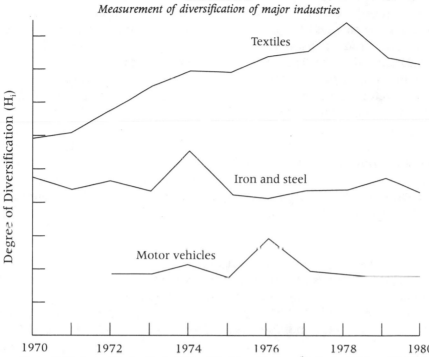

Source: F. Kodama, "Technological Diversification of Japanese Industry," *Science* 233 (July 18, 1986): 293. © AAAS.

system is built into management practice and, therefore, managers are motivated to diversify to keep employment constant.

Industry-by-Industry Comparison

Because we have defined an industrial sector's technological diversification as the sector's R&D activity outside its principal product fields, principal product fields must be distinguished from those that are not. The Statistics Bureau's survey, however, does not disaggregate company's R&D expenditures industry by industry, even though the expenditures are disaggregated by product field. Instead, expenditures are assigned to the company's primary industry, but many Japanese companies operate in several industries. Therefore, in order to disaggregate a company's R&D expenses industry by industry, we grouped the twenty-five industrial sectors in the survey into twenty-one sectors. Electrical machinery, electrical equipment and supplies manufacturing, and communication/electronic equipment manufacturing, are aggregated into the one sector, electrical machinery manufacturing. Industrial chemicals manufacturing, oil and paint manufacturing, and other chemical product manufacturing, are aggregated into chemical products manufacturing. There was one exception to this process. Transportation equipment manufacturing is disaggregated into motor vehicle manufacturing and other transportation equipment manufacturing to allow for the inclusion of shipbuilding. Thereafter, we called the sector shipbuilding instead of other transportation equipment manufacturing. Transportation, communication, and public utility industries are excluded from our data, because they are service industries and cover many different companies from railroads through electric utilities.

Once we had reorganized the industrial sectors, all the product fields were classified as principal product fields and assigned to the twenty-one industrial sectors, as shown in Table 2-1. Because almost all the names of product fields are identical to the names of industrial sectors, this classification was not difficult.

Table 2-2 shows each sector's R&D expenditure outside its principal product fields (denoted by B) and the relative size of the expenditure in terms of its ratio to total expenditure (denoted by A). In 1980, the amount of R&D expenditure outside principal product fields was ¥447 billion for all the industries. This corresponds to nineteen percent of their total R&D expenditure. The industries in which expenditures outside principal product fields exceeded expenditures for principal product fields (B/A over 0.5) are mining, textiles, fabricated metal products, and shipbuilding. The industries in which both types of expenditures are comparable are petroleum/coal, ceramics, and nonferrous metals. The industries in which the expenditures outside principal product fields are minimal compared to those for principal product fields (B/A below 0.1) are agriculture, drugs/medicines, and motor vehicles.

The degree of diversification in major industries after 1980 is shown in Figure 2-6. Diversification in the textile industry reached as high as 70 percent although the rate of diversification has saturated. On the other hand, the diversification in motor vehicles manufacturing remains quite low, and it will probably not change drastically in the near future. The industry that has changed most in terms of diversification is the iron and steel industry. Its degree of diversification, which was only 20 percent during the 1970s, has now reached almost 50 percent. Diversification may have been delayed in the iron and steel industry because of its monotechnology culture. The industry is a typical mass-production industry and its technology is based on a well-established

TABLE 2-2
Measurement of diversification, 1980

Industrial Sector	Total R&D Expense (¥ billion) A	Expense Outside Principal Products (¥ billion) B	B/A
Agriculture	2.2	0.10	0.04
Mining	8.9	6.25	0.70
Construction	53.0	9.55	0.18
Food	59.6	19.34	0.32
Textiles	17.9	9.64	0.54
Pulp and paper	13.1	2.42	0.19
Printing and publishing	4.8	1.24	0.26
Chemical products	285.3	45.58	0.16
Drugs and medicines	162.3	10.87	0.07
Petroleum and coal	24.7	11.28	0.46
Rubber products	42.9	6.33	0.15
Ceramics	62.1	26.18	0.42
Iron and steel	188.2	26.23	0.22
Nonferrous metals	37.5	15.48	0.41
Fabricated metals	38.4	20.36	0.53
Ordinary machinery	154.3	37.60	0.24
Electrical machinery	666.3	74.55	0.11
Motor vehicles	370.0	27.47	0.07
Shipbuilding	71.1	47.32	0.67
Precision equipment	66.4	19.34	0.29
Other manufacturing	60.4	30.40	0.50
TOTAL	2319.4	447.16	0.19

manufacturing process with few variations available. Therefore, we can assume that the monocultural aspects of corporate and technology strategy lessened diversification efforts compared to the diversification efforts of other declining industries. Such a finding would seem to confirm our hypothesis that diversification is used as a survival strategy for a mature industry.

SECTORAL PATTERNS OF DIVERSIFICATION

Industries can diversify their R&D activities in a number of directions. To identify these directions in relation to industrial structure, we need to see how the industries are interrelated. One interrelation that can be identified empirically is input to output.

FIGURE 2-6
Ratio of R&D expenditure in product fields outside principal fields to total R&D expense,
1975–1986

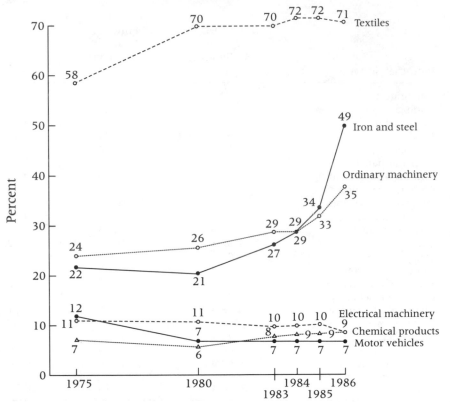

Source: Prime Minister's Office of Japan, Statistical Bureau, "Survey of Research and Development" (in Japanese) (Tokyo: Nihon-Toukei-Kyoukai, 1976–1987).

If an industry diversifies its R&D activities into product fields that are either inputs or outputs of that industry, the direction of diversification is *vertical*. If an industry does not diversify through input-output relations, the direction is *horizontal*. Materials industries, such as chemical manufacturing, are diversifying along the path of the sciences that they have mastered. Because this type of diversification is independent of the input-output relation, it is categorized as horizontal diversification.

There are two types of vertical diversification: *upstream* and *downstream*. If an industry's R&D activities are diversified into product fields that are inputs, the direction of diversification is *upstream*. A company that diversifies into the business area of its parts suppliers or into the business area of its equipment suppliers is diversifying upstream, although input-output statistics, which trace the commodity flow among different industrial sectors, may not reflect the direction of the diversification.

If an industry's R&D activities are diversified into outputs, the direction is *downstream*. More and more Japanese chip manufacturers, for example, are producing a variety of office machines because these machines are now electronic rather than mechanical products. Many Japanese camera manufacturers have also become active in office automation by extending their competence in lens manufacturing into business areas that are related to image-processing technology. These types of business expansion are categorized as downstream diversification.

Measurement of Direction

The measurement problem can be formulated as follows: To what extent does the direction of R&D investment, given by the distribution of R&D expense into product fields, follow upstream or downstream (given by the input-output transaction table)?

Given E_{ij} (*i-th* industry's expense into *j-th* industry's principal product fields, for $i, j = 1, 2, \ldots, n$), and letting $E_{ii} = 0$ because we are interested in diversification, i.e., expenses outside principal product fields, then, the *i-th* industry's direction of R&D investment can be described by the following unit vector:

$$\mathbf{r}_i = [r_{i1}, r_{i2}, \ldots, r_{in}],$$

where

$$r_{ij} = E_{ij}/(E_{i1}^2 + E_{i2}^2 + \ldots + E_{in}^2)^{1/2} = E_{ij}\bigg/\sqrt{\sum_j E_{ij}^2},$$

so that the *norm* of $\mathbf{r}_i$ equals 1, i.e.,

$$|\mathbf{r}_i| = \sqrt{r_{i1}^2 + r_{i2}^2 + \ldots + r_{in}^2} = 1.$$

Given T_{ij} (the amount of transaction from i-th industry to j-th industry) and letting $T_{ii} = 0$, then, the i-th industry's upstream direction is represented by the unit vector

$$\mathbf{u}_i = [u_{1i}, u_{2i}, \ldots, u_{ni}],$$

where

$$u_{ij} = T_{ij} \Big/ \sqrt{\sum_i T_{ij}^2}, \text{ thus } |\mathbf{u}_i| = 1.$$

Now, we can try to develop an indicator of the i-th industry's upstream diversification. The indicator will be denoted by U_i, and should be designed to satisfy the condition: the higher the U_i value, the closer the direction of the i-th industry's R&D investment to the i-th industry's upstream direction. Let the difference in direction between $\mathbf{r}_i$ and $\mathbf{u}_i$ be θ_u, then, $cos\ \theta_u$ can be expressed in terms of the inner product of the two vectors:

$$\cos \theta_u = \mathbf{r}_i \cdot \mathbf{u}_i = \sum_k r_{ik} \cdot u_{ki}.$$

Therefore, U_i can be developed and measured by

$$U_i = r_{i1} \cdot u_{1i} + r_{i2} \cdot u_{2i} + \ldots + r_{in} \cdot u_{ni}.$$

Similarly, the i-th industry's downstream direction is represented by the unit vector

$$\mathbf{d}_i = [d_{i1}, d_{i2}, \ldots, d_{in}],$$

where

$$d_{ij} = T_{ij} \Big/ \sqrt{\sum_j T_{ij}^2}, \text{ thus } |\mathbf{d}_i| = 1.$$

An indicator of downstream diversification of i-th industry, D_i, can then be developed and measured by

$$D_i = \boldsymbol{r}_i \cdot \boldsymbol{d}_i = r_{i1} \cdot d_{i1} + r_{i2} \cdot d_{i2} + \ldots + r_{in} \cdot d_{in}.$$

TABLE 2-3
Measurement of direction of diversification, 1980

Industrial Sector	Degree of Diversification (B/A)	Direction of Diversification	
		Upstream (U_i)	Downstream (D_i)
Agriculture	0.04	0.396	0.811
Mining	0.70	0.032	0.033
Construction	0.18	0.042	0.059
Food	0.32	0.055	0.099
Textiles	0.54	0.328	0.026
Pulp and paper	0.19	0.203	0.137
Printing and publishing	0.26	0.047	0.057
Chemical products	0.16	0.026	0.058
Drugs and medicines	0.07	0.235	0.205
Petroleum and coal	0.46	0.035	0.154
Rubber products	0.15	0.136	0.424
Ceramics	0.42	0.058	0.073
Iron and steel	0.22	0.032	0.179
Nonferrous metals	0.41	0.031	0.146
Fabricated metals	0.53	0.024	0.079
Ordinary machinery	0.24	0.119	0.111
Electrical machinery	0.11	0.063	0.243
Motor vehicles	0.07	0.083	0.118
Shipbuilding	0.67	0.274	0.010
Precision equipment	0.29	0.074	0.247
Other manufacturing	0.50	0.402	0.075

Table 2-3 gives the calculated values for the degree of upstream diversification and downstream diversification.[10] Among the relatively diversified industries (*B/A* higher than 0.19), the industries clearly following the path of upstream diversification are textiles, pulp and paper products, and shipbuilding. The industries clearly following the path of downstream diversification are petroleum and coal products, iron and steel, nonferrous metals, and precision equipment. There are some industries—mining, printing and publishing, ceramics, and fabricated metal products—that show neither upstream nor downstream diversification. Because they are diversified, however, we can assume they follow the path of horizontal diversification.

Technological diversification does not always lead to product diversification. An industry that is upstream in relation to the majority of industries can make a technological diversification further upstream. The iron and steel industry, for example, can diversify further upstream by engaging in engineering. To do so, the industry has to diversify its

R&D activities into the product fields of its equipment and machinery suppliers.

Other possible upstream product fields for the iron and steel industry include fields related to the design of the industry's factories. Indeed, excluding those from engineering and construction companies, the active membership of the Engineering Advancement Association of Japan (ENAA) is composed of individuals from the major iron and steel and shipbuilding companies. These two industries are utilizing the experiences that have been accumulated in designing factory automation in their principal product areas to diversify further upstream. In fact, both industries have successfully transformed previously disjointed processes into continuous, computer-controlled processes.

Identification of Sectoral Patterns

Many industries, such as drugs and medicines, have diversified both upstream and downstream. Although U_i can be compared to U_i or D_i can be compared to D_i, U_i cannot be compared to D_i. To make them comparable and discover the sectoral pattern, U_i and D_i must be normalized. The normalized value of U and D can be calculated by

$$N_u = (U - M_u)/S_u, \text{ and } N_d = (D - M_d)/S_d,$$

where M_u and M_d are the averages of U and D, respectively, and where S_u and S_d are the standard deviations of U and D, respectively.

By plotting (N_u, N_d) within the x-y plane, we can classify sectoral patterns of technological diversification into four categories, as shown in Figure 2-7. Quadrant I includes industries that have both strong upstream and downstream diversification, such as drugs and medicines. The sectoral pattern is identified as vertical diversification. Quadrant II includes industries that have strong downstream diversification (D_i is above average) but weak upstream diversification (U_i is below the average). Therefore, the sectoral pattern in this quadrant is identified as downstream diversification. Quadrant III includes industries that have both weak upstream and downstream diversification and therefore are supposed to have strong horizontal diversification. Quadrant IV includes industries that have strong upstream but weak downstream diversification. The pattern in this quadrant is positively identified as upstream diversification.

The industries that can be clearly classified as diversifying upstream are textiles and shipbuilding. These are more or less mature industries

FIGURE 2-7
Identification of sectoral patterns

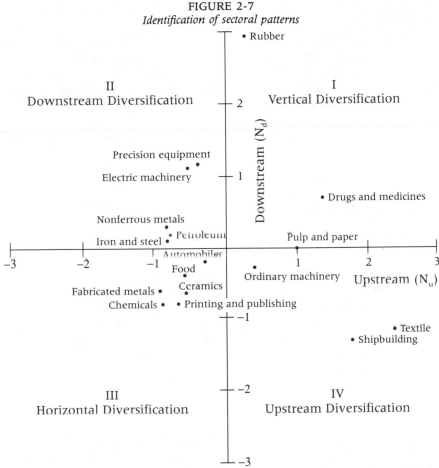

Source: F. Kodama, "Technological Diversification of Japanese Industry," *Science* 233 (July 18, 1986): 294. © AAAS.

in the sense that industries in newly industrialized economies (NIEs) are catching up with them. Therefore, these industries are diversifying upstream in order to maintain their international competitiveness. In the next section I will argue why this upstream diversification is rational behavior.

Industries classified as diversifying downstream include electrical machinery and precision equipment. These are typical high-tech industries that grew rapidly in the period following the first oil crisis. The revolution in electronics widened applications for these industries and induced companies to diversify downstream. We can hypothesize that drastic growth of an industry can be fostered only by downstream diversification.

Those industries that show horizontal diversification include chemical products and fabricated metal products. Since they are typical material industries, their diversification is rather free from industrial structural relations. Finally, the drug and medicine industry, where the introduction of biotechnology is supposed to be promising, is identified as showing vertical diversification, i.e., both upstream and downstream. The difference in the diversification patterns of chemical products and pharmaceuticals will be discussed in the next section.

TESTING HYPOTHESES

Having measured the direction of diversification and identified the direction for each industrial sector, we can now test several hypotheses concerning the diversification behaviors of industries involved in high-tech development. We can also test several theories on technological diversification patterns that were developed by American scholars and based on the past experiences of U.S. companies. It may be that Japanese experiences are different or that the emergence of high technology is allowing Japanese industries to follow trajectories that are different from the trajectories derived from these theories.

Trickle-Up Growth

We have hypothesized that the drastic growth of an industry can be fostered only by downstream diversification, and we have stated that one of the trajectories for downstream diversification is the result of a trickle-up process, i.e., a localized technical change produced by a component supplier will gradually find wider applications in downstream product areas.

The trickle-up diversification hypothesis, as we can call this hypothesis, then, states that downstream diversification through the trickle-up process will result in business growth. We have already measured the degree of downstream diversification for 1980, and we can use the production index of 1985, with 1975 as a base year, to represent business growth. Table 2-4 gives these two variables for all the industries.

Sixteen industries from Table 2-4 were selected for a statistical test. A regression analysis between D_i (the degree of downstream diversification measured in 1980) and P_i (the production index of 1985 with 1975 as a base year), yields the following equation:

TABLE 2-4
Downstream diversification and production growth

Industrial Sector	Downstream Diversification (1980)	Growth in Production (1985/1975)	Identification
Food	0.099	1.15	
Textiles	0.026	1.06	
Pulp and paper	0.137	1.50	
Printing and publishing	0.057	—	
Chemical products	0.058	1.40	
Drugs and medicines	0.205	2.71	high-tech
Petroleum and coal	0.154	0.85	
Rubber products	0.424	1.61	irregular
Ceramics	0.073	1.27	
Iron and steel	0.179	1.26	
Nonferrous metals	0.116	1.13	
Fabricated metals	0.079	1.32	
Ordinary machinery	0.111	1.91	high tech
Electrical machinery	0.243	4.15	high-tech
Motor vehicles	0.118	1.87	
Shipbuilding	0.010	0.56	
Precision equipment	0.247	3.80	high-tech

Source: Production index from the Ministry of International Trade and Industry, weighted by value added. MITI, *Industrial Statistics* (Tokyo: Nihon-Toukei-Kyoukai, 1990).

$$\log P = 2.80 \cdot D + 0.02, \qquad R^2 = 0.311$$
$$(2.52)$$

The coefficient of determination is so low that the trickle-up diversification hypothesis is not valid statistically. In fact, it is hard to find a solid relationship between these variables. A closer look at Table 2-4, however, reveals that the low coefficient of determination may be due to a specific irregularity: the degree of downstream diversification for rubber products is extremely high compared to the degrees for other industries.[11] Because we are analyzing high-tech industries, there is no positive reason to keep this industry in the data set. Therefore, we can delete this industry and try another regression analysis, using fifteen industries. When this is done, the scatter diagram between the two variables becomes much more organized, as depicted in Figure 2-8. As shown in the figure, the two variables are fairly correlated with each other. A regression analysis, however, yields the following result:

$$\log P = 5.80 \cdot D - 0.31, \qquad R^2 = 0.624$$
$$(4.64)$$

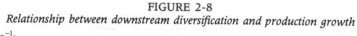

FIGURE 2-8

Relationship between downstream diversification and production growth

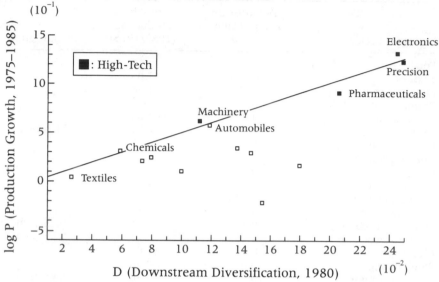

The coefficient of determination is not high enough to validate the causality of trickle-up diversification when we look at all the manufacturing industries.

This result, however, may give us a chance to confirm our hypothesis that trickle-up diversification is unique to high-tech industries. To do so, we need to confine the test to those industries that will be identified as high-tech industries in chapter 3—drugs and medicines, ordinary machinery, electrical machinery, and precision equipment. In the scatter diagram of Figure 2-8, these industries are represented by black squares. A regression analysis using these four samples yields

$$\log P = 5.28 \cdot D + 0.04, \qquad R^2 = 0.925$$
$$(4.96)$$

Because we have obtained a very high coefficient of determination, our hypothesis seems to be true statistically. Among high-tech industries, therefore, the larger an industry's degree of downstream diversification, the higher the industry's production growth. Thus, trickle-up diversification is the business growth strategy for high-tech industries.

Moreover, we can interpret the regression line shown in Figure 2-8 in a broader theoretical framework of business growth. That is, we can say the regression line composes the *frontier* of growth potential achievable for all the industries studied. However, it is only the use of trickle-up strategy that makes it possible to exploit this potential. Therefore, trickle-up diversification is a method of attaining the maximum potential level of business growth.

The frontier line suggests that other industries could have exploited the potential growth available through downstream diversification if they had followed the trickle-up trajectory. In fact, there are two industries—chemicals and textiles—that are on the extrapolated frontier line. In the chemical industry, we know that Toray followed the trickle-up approach in developing carbon fiber and achieved maximum growth. In textile industry, Asahi-Kasei Co. implemented the trickle-up strategy by applying its fiber technology to filters for kidney dialysis machines and building materials.

It should be noted that the automobile industry is close to the frontier line. In fact, the business strategy of automobile industry in Japan is often characterized by short product cycles, continuous product improvement, common component parts among different car models, overlapped product development cycles, and lean production.[12] It would seem then that these characteristics are related to the trickle-up trajectory.

Product-Cycle Theory

Although Vernon's product-cycle theory implies abandoning a mature technology to stay ahead of competitors, Japanese industry is not following this strategy. The emergence of high technologies seems to be making another strategy available. Upstream diversification seems to allow a mature industry to keep its competitive edge against NIEs by compensating for the loss in final product market through the exportation products that are its industrial inputs and through plant engineering. This leads to the hypothesis that an industry's loss in international competitiveness will lead to upstream diversification.[13] To test this hypothesis, we must compare an industry's international competitiveness, as measured by its export growth, with its degree of upstream diversification (see Table 2-5). Because we are concerned with the behavior pattern of matured industries, only industries with export growth rates below average (the average annual growth rate of all the Japanese industries was 1.22 percent in the 1970s) were selected. From these industries with products that are not greatly exported (those with an export ratio to domestic production of less than three percent), were excluded. This left seven industries for our sample.

We can try to explain their degree of upstream diversification (U_i)

TABLE 2-5
Export growth and upstream diversification

Industrial Sector	Export Growth in 1970–1980 (% per year)	Upstream Diversification (in 1980)	Export Ratio (%)
Food	1.09	0.055	1.39
Textiles	1.02	0.328	9.24
Pulp and paper	1.17	0.203	2.55
Printing and publishing	—	0.047	0.65
Chemical products	1.18	0.026	11.53
Drugs and medicines	1.16	0.235	1.74
Petroleum and coal	1.27	0.035	2.23
Rubber products	1.23	0.136	15.32
Ceramics	1.17	0.058	5.24
Iron and steel	1.18	0.032	11.75
Nonferrous metals	1.22	0.031	7.42
Fabricated metals	1.19	0.024	8.49
Ordinary machinery	1.24	0.119	13.36
Electrical machinery	1.26	0.063	17.16
Motor vehicles	1.88	0.083	18.26
Shipbuilding	1.13	0.274	42.35
Precision equipment	1.11	0.074	28.82

Source: Export statistics from United Nations, *World Trade Statistics* (New York: United Nations, 1985); export ratio from Ministry of International Trade and Industry, *Industrial Statistics* (Tokyo: Nihon-Toukei-Kyoukai, 1985).

measured in 1980, by means of their average annual export growth rate (E_i) from 1970 to 1980, as shown in Figure 2-9. The regression equation is

$$U = -1.80 \cdot E + 2.20, \qquad R^2 = 0.808.$$
$$(4.58)$$

With a fairly high coefficient of determination, upstream diversification is negatively related to export growth; that is, the lower an industry's export growth rate, the larger the industry's degree of upstream diversification. Therefore, the upstream diversification strategy for mature Japanese industries is valid statistically. The negative relationship between export growth and upstream diversification provides us with a scenario in which the loss in market does not lead to a protectionist movement. With upstream diversification, employment can be maintained at individual firms. This would appear to be the strategy followed by mature Japanese industries, such as textile, shipbuilding, and iron and steel.

FIGURE 2-9

Relationship between export growth and upstream diversification

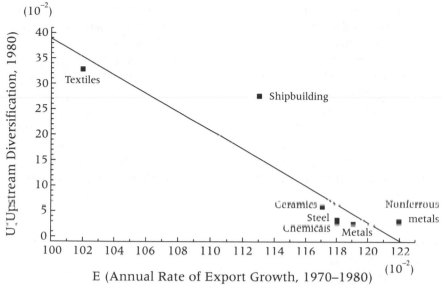

This negative relationship also implies that mature industries can take advantage of the technological continuity that is present in all industries in this high-tech era, even when the trickle-up strategy is not available to them. Instead of abandoning an industry's core competence for new technologies, the industry can diversify its business along its industrial technological linkages. In fact, the emergence of high technologies is making this type of business diversification more plausible and viable than ever before.

Japanese Management

Although chemical products manufacturing and electrical machinery manufacturing are often said to be similar because both are highly dependent on scientific findings,[14] these industries are different in their technological diversification: electrical machinery diversifies downstream, while chemical products diversifies horizontally.

Although both industries are heavily dependent upon basic science, research in the electronics industry has a clear-cut direction, whereas research in the chemical industry is undirected and requires broad-based support. In the electronics industry, therefore, management can implement an innovation cycle from basic research to commercialization using a team approach. In the chemical industry, managing basic research is difficult because managers must rely on the contributions from a few talented individuals. This difference in diversification patterns explains the widely accepted observation that Japan has a strong electronics industry but not a strong chemical industry.

The pharmaceutical industry, which used to be similar to the chemical industry, has changed because of the emergence of biotechnology. The direction of diversification in the pharmaceutical industry has now become vertical, while the chemical industry still follows horizontal diversification. This may confirm our finding that diversification in biotechnology follows a trickle-up trajectory, and it may be one reason why it has been suggested that Japan will become the leading competitor of the United States in pharmaceuticals.[15] Moreover, if our characterization of the diversification pattern for the chemical industry is accurate, it is unlikely to be accurate very long. As we noted earlier, the development pattern of new materials is following the trickle-up trajectory.

Government Policy

In Japan, research associations are widely used to implement government R&D policies.[16] The research associations, which are of limited duration, are made up of competing firms that share researchers and costs. The government subsidizes the research through funds and tax benefits for the associations. This approach was adapted from the British model and reformulated as the engineering research association (ERA, *Kenkyu-Kumiai*) in 1966.

There are various built-in mechanisms to accelerate the technological diversification of firms participating in these associations. One mechanism is *venturing*. Although venturing is common and supported by government subsidies in the United States, ventures are not in-house.[17] In Japan, on the other hand, ventures are usually in-house. Not too long ago, we suggested that ERAs were being used by big Japanese firms as springboards into in-house ventures. It is standard practice for firms participating in an ERA to set up an in-house project team that has roughly the same number of members as the research team that the company sends to the ERA. The project team supports its colleagues on assignment and assimilates the data generated by the ERA. In this way, the project team is an in-house venture unit. When the ERA disbands and the employees on assignment return to the company, they add their weight to the project team. Thus, the team serves, in effect, as the headquarters of a venture-capital business.[18]

Indeed, the choice of an ERA research project is often related to the product fields that are considered marginal by participating firms so that it causes less conflict of interest among them.[19]

MANAGING TRICKLE-UP DIVERSIFICATION

There has been a tendency to assume that Japanese business managers are less risk-averse than their American counterparts. However, the management approach that we have observed in trickle-up diversification is evidence of a determination to manage technical risk in a conservative, well-organized way. Based on our observations, then, trickle-up diversification is a rational approach to the management of technology. First, it divides technical and market risks into smaller elements, through short product cycles and quick response to feedback from the market. Second, it pays attention to realistic manufacturing practices as a means of reducing technical risk. Priority is given to understanding production process technologies as well as to product design and function prior to a commitment to commercialization. Third, it makes possible the continuing search for signs of emergent technologies outside the firm and the industry that might threaten the attractiveness of strategically important markets. Finally, it promotes internal diversification as a response to those early warnings.[20]

The trickle-up approach can be linked to the notion of core competencies, which, as we noted earlier, is defined as "collective learning in the organization." Recent studies have focused on core competencies as key elements to a firm's success and have emphasized the interplay of core technologies and product strategies.[21] Roberts and Meyer studied correlations between the extent to which firms focus their technologies to produce "core products" and their success in sustaining dynamically changing businesses.[22]

Meyer and Utterback carried the analysis further in 1992 with an empirical analysis of the relationship between diversity of product "cores" and business success. By mapping the history of both product and market novelty in the product portfolios of individual firms, they distinguished "focused" from "unfocused" firms. Among other things, they concluded that

> A corporate product technology strategy in which each new product concept is an *extension* of a well-developed existing and successful *product core* (the Double Header pattern) will be more highly successful than will one

in which each new product concept is based on a new or emerging product core (the Home Run Hitter pattern).[23]

We can interpret the trickle-up approach as an effective and inevitable way of extending a core competence. It is, then, a business diversification strategy that should be taken by high-tech firms in a dynamically changing business environment.

Typology for Business Strategy

If the trickle-up trajectory is the business diversification pattern of high-tech firms, how can a firm extend its core competence beyond its current business scope? Is there any consistent business strategy that makes it possible for a firm to follow the trickle-up trajectory?

As we have already noted, every company must design a business strategy to support not only the business environment (financial resources, regulatory constraints, human resources, and so forth) but also its market, its technology, its products, and its customer relationships. When a company establishes its strategy and organizes the management of its technical activities, it may weigh these variables equally, or it may weigh one of them more heavily than the others.

A firm that optimizes its business around a defined market is likely to diversify its technology to guard against unexpected competition from technology not usually found in its market. When a firm's business strategy is largely defined by its technology, the firm may choose to constrain the scope of its technologies to ensure a leadership position and will map this advantage onto many different markets. A firm optimizing its business around a set of selected products may stress the specific set of technologies necessary for those products and may specialize in its ability to switch from one technology to another. A firm characterized by responsiveness to one or only a few large customers will be expected to develop the requirements for a technology with its customer.

The Market-Focused Strategy.   Some firms develop their business strategy for customers in broad, functionally defined markets. These companies will move into any technological area required to increase or preserve their share of the defined market.

SEI offers the best example of this business strategy. For years, SEI has defined itself as a supplier of communication cable. It defines its core competency as the manufacture of cables for communications. Its diversification strategy, therefore, is formulated around this competence.

In the past, SEI specialized in the development of copper cables that were used to create worldwide telephone and coaxial cable networks. Planning for future expansion, SEI's management carefully studied trends in their self-defined market and saw computer and digital telecommunications as the most rapidly growing areas of information communication. Technological improvements and changing customer demands eventually led to a reformulation of SEI's business strategy, which then focused on fiber optics. When it entered the fiber optics market, SEI successfully diversified internally. To pursue a market-focused business strategy, SEI relies on internal diversification to defend its business model (cables for communications) against new technologies (such as optical communications) that affect the market for information transmission.

Figure 2-10 shows the growth and trickle-up diversification of SEI, which were made possible by Sumitomo's R&D laboratories. In the late nineteenth century, the company's core business was copper mining and smelting. During the early part of the twentieth century, SEI used its skills in smelting to move its business focus into copper wire manufacturing. In the immediate postwar era, the company drew upon its technological base in metals and wire to move into special steel wires, its base in wire coatings to move into rubber and plastic products, and its base in electronics to move into electronic materials and antenna systems. By the 1960s, the company had moved into more complex systems technologies such as integrated electronics systems and disc brakes. In the 1970s and 1980s, Sumitomo used these built-up technological competencies to move into high-technology electronic systems (workstations) and automotive electrical systems.

To realize profitable businesses in such diversified technologies, SEI devised a thirty-year plan. The first ten years called for SEI to develop and apply the new technology. The second ten years would be spent entering the market and then gaining market share. Only during the last ten years, would the technology be expected to earn sufficient profits to pay back the investment. It takes a strong strategic commitment to sustain internal diversification over such a period of time.

For a market-focused strategy using a trickle-up trajectory to succeed, a firm has to be able to use, at each stage of its development, both new and existing R&D resources to develop new commercial technologies and enter new business areas. The company needs to be particularly adept at using its existing technological capabilities and competencies to enter new technology-intensive, high-growth fields.

The Technology-Focused Strategy. Some companies develop their competitive advantage through selected areas of technical know-how,

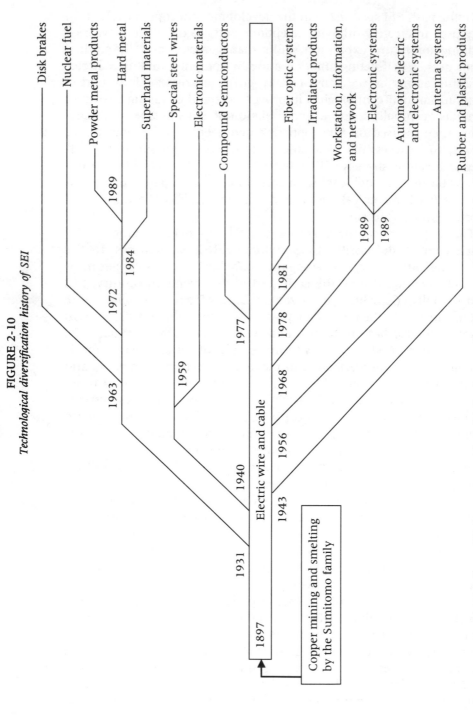

FIGURE 2-10
Technological diversification history of SEI

Source: Tsuneo Nakahara, SEI (paper presented at the Seminar on the Development of High Technologies in the U.S.A. and Japan: Comparative Evaluation and Policy Implications, John F. Kennedy School of Government, Harvard University, 1992). Reprinted by permission.

processes, and facilities. Unlike market-focused businesses, which will move into any technological area required to increase or preserve their market share, these companies seek to develop expertise in a few core technologies that are strategic to the business of the company. A company defined by its technology will search for the minimum number of technologies that support a spectrum of growth markets, will be willing to drop some markets and pick up new ones, and will make sure that its technology is in a leadership position.

NEC Corporation is perhaps the best example of a firm with a technology-focused business strategy. The trickle-up process implemented by this strategy can be illustrated by the development of personal computers in Japan, in which NEC played a pivotal role.

In Japan personal computers underwent major changes while adapting to market needs. The development of microprocessors, which provided the technological basis for personal computers, grew out of the progress made in pocket calculators in the 1960s. The 4-bit microcomputer 4004, the first general-purpose microprocessor developed by Intel Corporation in the United States, was born from the development of an LSI circuit requested by Busicon (Business Computer Inc.), Japan's pocket calculator manufacturer. To use this LSI circuit for various applications, Busicon asked Intel for a programmable circuit configuration. To meet Busicon's request, Intel incorporated the architecture of a mainframe computer into the LSI circuit, resulting in the development of a single-chip microprocessor.

This was followed by the development of the 8-bit microprocessor 8008 for use in computer terminals capable of handling alpha-numeric character codes. That microprocessor became the basis of personal computers. Along with the development of these microprocessors, 1K-bit memories (DRAM, ROM) were developed. At this stage, it became technologically possible to produce compact, low-cost computers.

Although venture businesses like Apple opened the personal computer market in the United States, in Japan, the personal computer emerged from training kits produced by semiconductor manufacturers that wanted to find outlets for their products. As the demand for personal computers for businesses grew, large companies previously engaged in the manufacture of mainframe computers made inroads into the market, expanding their shares rapidly. Within NEC, the initiative was first taken by the device division and then passed to the information-processing division.

Although manufacturers did not expect much demand for the microcomputer kits, NEC's creation of Bit-Inn Centers for user services and technical support, coupled with some other factors, brought about a boom in sales. The use of training kits expanded, and gradual, yet steady, advances were made toward higher levels of use. All these developments finally crystallized in the advent of a personal computer as a package system, containing both the CPU and peripheral equipment.

When NEC's PC-8001, the first best-selling personal computer made in Japan, reached shipment levels of 200,000 units, software development began to grow. As game software was expanding into the consumer market, a wide range of word-processing and other business applications were developed. The development of 16-bit microprocessors made business machines increasingly sophisticated and hobby machines easier to use, as well as more affordable, through unification of standards centering on MSX. In the meantime, specialization of specific functions brought about two new products: word processors and family computers for game use.

It is important to realize that a firm following the technology-focused strategy does not derive its products from innovations in technology. Instead, the firm is heavily engaged in testing many markets for opportunities, looking for new needs that can be filled by its capabilities. Core technologies are defined through a planning process that minimizes technological diversity in order to maximize technological competitive advantage. The markets to be served are the independent variables; their selection and management is therefore a matter of constant review and choice.

Product-Focused Strategy.   Some firms pursue a discrete set of products that have functional attributes defined from the perspective of the end-users. These products may be sold to a wide range of users. Unlike market-focused strategy, which hopes to penetrate all segments of a market, product-focused strategy concentrates on specific groups of products that do not have a lot in common. Unlike the technology-focused strategy of some firms, in which products are derived from a base of core technologies, firms using product-focused strategies create products that incorporate whatever technologies are required, whether or not these technologies are available in-house.

Sony Corporation best represents a product-focused approach, although it could be argued that Sony is a market-focused consumer electronics company. If Sony's business strategy were primarily market-focused, however, its business objectives would be to penetrate all segments of the consumer electronics market. In actuality, Sony's focus is on specific groups of products that do not have a lot in common: the high end of the TV, VCR, and display industry; miniaturized, high-quality audio products for consumers; and the electronic camera market.

Described in this way, Sony's approach may not seem to follow the trickle-up trajectory. When we review Sony's history of major product innovations, however, we find that Sony has been following the trickle-up trajectory toward higher-ends both in technology and in the market. In 1955, after four years of R&D investment, Sony developed its first transistorized radio. It was followed by the eight-year development of the Trinitron television picture tube. In 1983, after thirteen years of R&D investment in charge-coupled device (CCD) technology, Sony developed an electronic camera. Although each development put a significant strain on the finances of Sony at the time, a corporate product vision drove the development of a specific technology. Still, Sony did not attempt to cover all the technologies common to the consumer electronics industry.

At present, Sony is moving into businesses that place a premium on delivering complex systems of hardware and software to sophisticated business customers. Its success with CCD imaging technology has given it a commanding position in television production equipment as well as in consumer video cameras. If Sony's technology can begin to compete with photographic film in the motion picture industry, another large market will be open to it. Video tape, video disks, and multimedia computer terminals, all require video software. Sony not only provides that software, it has entered the motion picture production business through its acquisition of CBS and Columbia Pictures.

Sony prides itself on its ability to move quickly into any technology needed for a strategic product. It defines itself by its technological agility, not by core technologies. Therefore, Sony's technology strategy is product driven and focuses on the ability to master or acquire any technology needed to realize its product goals. This product-driven strategy, however, does create multiple-product core competencies at a lower level of aggregation, as illustrated in Figure 2-11, which shows the competencies built in the case of CD technologies. One must be careful about how one defines core competencies that support corporate product strategy. Sony's core competence, as we have noted above, is its ability to create and acquire technologies on demand from key-product goals. Such a competence is impressive from the perspective of technology management.

The Customer-Relationship-Focused Strategy. There are companies that deal with a few customers in a monopolized market. For example, software and hardware suppliers for electric or gas utilities basically deal with a single customer.

FIGURE 2-11
CD and related technologies at Sony

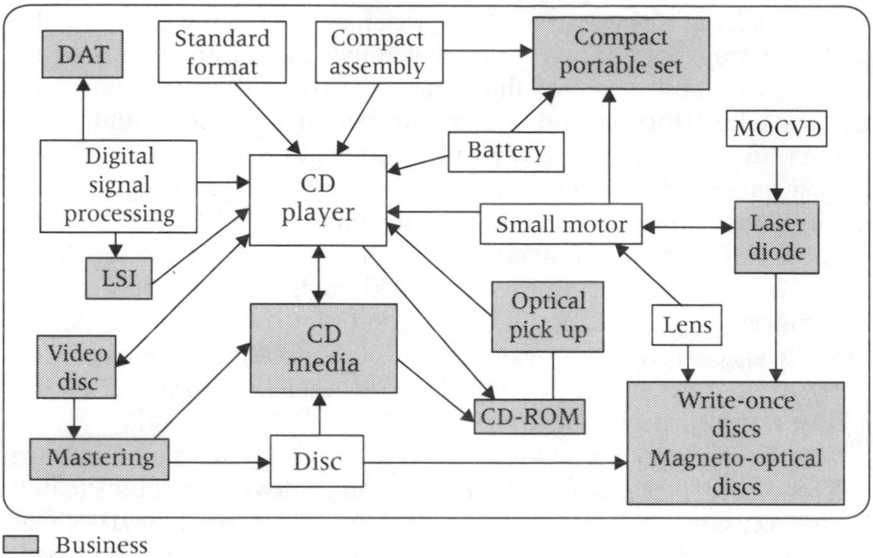

☐ Business

Source: Seiichi Watanabe, Sony (paper presented at the Seminar on the Development of High Technologies in the U.S.A. and Japan: Comparative Evaluation and Policy Implications, John F. Kennedy School of Government, Harvard University, 1992). Reprinted by permission.

Companies that pursue a customer-relationship-focused strategy are market-focused but in monopolistic markets. Therefore, they may develop their products in close cooperation with a single customer or a narrow spectrum of customers with similar functional requirements. These customers are concerned with how the technologies generated by their suppliers are integrated into systems that provide services to society. Because of their concerns, these firms may operate their own research laboratories and employ large numbers of scientists and engineers.

Japanese electric utility companies, for example, are very large and, unlike most of their U.S. counterparts, engage in a substantial amount of systems engineering research, often in collaboration with their suppliers, universities, and others. Tokyo Electric Power Company (TEPCO) is the largest private electric utility company in the world and one of only nine electric utility companies in Japan. It serves greater metropolitan Tokyo and supplies electricity to some 20 million customers. Its total annual sales are as great as 220 billion kilowatt-hours, about one-third of all the electric power consumed in Japan. In order to illustrate its size, we compared it with Boston Edison, as shown in

TABLE 2-6
Size comparison between Japanese and U.S. electric utilities

	Supply area (km²)	Population (million)	kWh Sales (billion)	Capital (¥ billion)	Employees
Japan Total	372,600	124.0	654	2,570	141,055
TEPCO	39,500	41.0	220	670	39,640
Boston Edison	1,500	1.5	14	120	4,738

Source: Tsuneo Mitsui, Tokyo Electric Power Company (paper presented at the Seminar on the Development of High Technologies in the U.S.A. and Japan: Comparative Evaluation and Policy Implications, John F. Kennedy School of Government, Harvard University, 1992).

Table 2-6. Because of its size alone, TEPCO is certain to play a large part in defining the technology provided to it, especially at the system level, by manufacturers, which include Toshiba and Hitachi.

There are, however, four other factors that strengthen TEPCO's role in defining the technology provided to it. First, of 40,000 employees, 25,000 are engineers and of those engineers, 400 are in research. Second, TEPCO collaborates with manufacturers and shares their risks in the development of major new technologies. At present, TEPCO is collaborating with GE, Hitachi, and Toshiba on a project for a new advanced design nuclear plant. Third, TEPCO sets extremely challenging requirements (the average duration of a power outage is only seven minutes compared with one to three hours in the United States, the United Kingdom, and France). Fourth, TEPCO invests in technologies, for example, electric automobiles, that meet both its sociopolitical interests and its business interests.

Mitsubishi Electric Corporation (MELCO) has employed the customer-relationship-focused strategy throughout its history. Its major customer base is the electric utility industry, in particular, Kansai Electric Power Company. In its early years, MELCO manufactured heavy electrical apparatus for electric utility companies. As electric power networks expanded, the need for network system analysis technology emerged. MELCO was one of the first companies to invest in this new technology. With the emergence of digital technology, MELCO's systems operation and planning business was born. It is now as large in overall sales as the heavy electrical apparatus business. With the advent of expert systems, the future will demand even greater innovative uses of information technologies.

MELCO's development of software for the electric utility industry best illustrates the trickle-up process in the customer-relationship-focused strategy. For MELCO and other companies with the same strat-

FIGURE 2-12
MELCO's concurrent engineering of software development for an electric utility

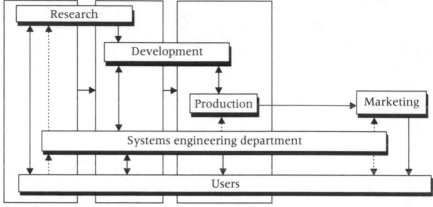

Basic Research   Prototyping   System Production

⟶ Technology and Product Flow
·····▶ Stakeholder Management

Source: Junichi Baba and Toshiaki Sakaguchi, MELCO (paper presented at the Seminar on the Development of High Technologies in the U.S.A. and Japan: Comparative Evaluation and Policy Implications, John F. Kennedy School of Government, Harvard University, 1992). Reprinted by permission.

egy, core competence evolves through interaction with its customer. This interaction is critical because the customer must be involved in defining system requirements and system engineering strategy. In addition, the customer has a diverse set of stakeholders whose interests must be considered: the public, the utility company engineers, producer company programmers, and government agencies.

Electric power systems technology has progressed through a series of distinct product categories (off-line systems evaluation, real-time open-loop control, power network management, and intelligent network control). Each product category requires its own approach to the innovation process. Increased demand and continuous development of a product can result in the creation of new product businesses. Customer evaluation of a new product prevents the uncontrolled creation of new products that are justified only by the producer's ideology.

MELCO has learned that the software development involves close collaboration between the user, the manufacturer, and the software development team. In this collaboration, known as "concurrent engineering," users play an important role in all stages of development, from prototyping to marketing, as shown in Figure 2-12. To foster concurrent engineering, MELCO established a systems engineering department that develops the architecture of the system and monitors the system's development, testing, and installation in close coordination with the department's counterparts in the customer's organization. Fifteen engineers in MELCO's department play a pivotal role in networking all the relevant players. The department does not do design work but identifies its customer's product needs and establishes the technical relationships and requirements with its customer's engineering groups. In the United States, an outside consulting firm may perform this work.

MELCO's trickle-up strategy will become more obvious when the company sees the possibility of diffusing its operational command and control software for power distribution networks beyond electric utility businesses. In the real-time computer control systems used in central power distribution stations, standardization, modulation, and decentralization have increased. It may be possible, therefore, to apply this technology to areas that are still more complex and demanding in terms of real-time control, such as road, railroad, and air traffic control systems.

Several types of expert systems now support power stations operators. By enhancing performance and reliability, it is possible that these systems can become fault-detection and diagnostic systems for equipment in the experimental module for a space station under development in Japan. Because operators at central power distribution stations have to perform individual jobs in a coordinated way, knowledge sharing is necessary and a man-machine interface is indispensable. Systems that permit this may be transferable to traffic control for space transportation. Real-time computer systems that simulate the power networks that are available in case of network failures are used to train operators. By enhancing the security level of these systems, they may be used for the control of gas and water in emergency situations, such as earthquakes. A system used for this, however, must deal with physical dimensions that are more demanding than those in electric power distribution in terms of real-time command and control hardware and software.

In the past, because of their budgets and technical sophistication, defense projects and other national science projects have been sources of spin-offs to civilian sectors. This transfer mechanism is becoming less effective, however, because of the spread of science and technology in general and the difficulty in concentrating monetary and human resources in a few specific national projects. On the other hand, Japanese electric utility companies clearly have a strong voice in the technology provided to them by their suppliers. We would argue, therefore,

FIGURE 2-13
Toshiba's spiral innovation process of LCD

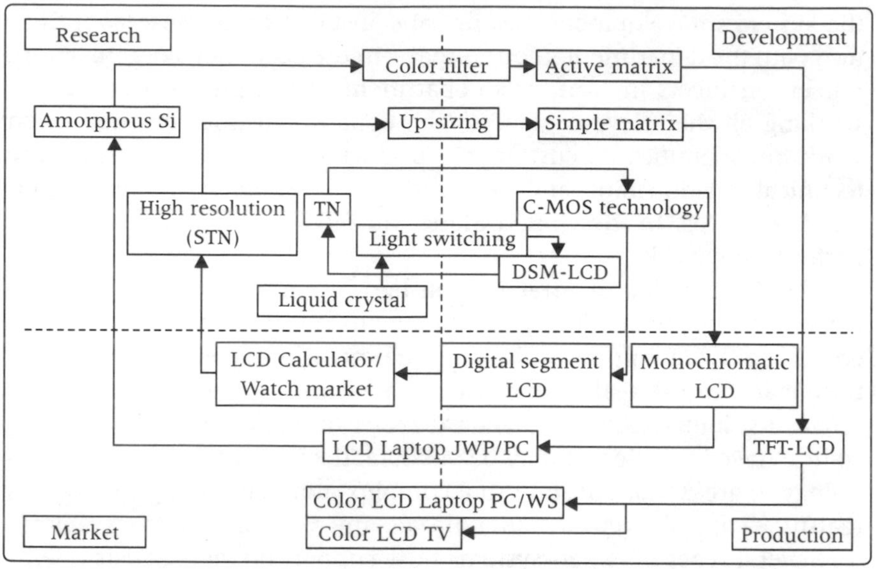

Source: Sei-ichi Takayanagi, Toshiba (paper presented at the Seminar on the Development of High Technologies in the U.S.A. and Japan: Comparative Evaluation and Policy Implications, John F. Kennedy School of Government, Harvard University, 1992). Reprinted by permission.

that Japanese electric utility companies may become major sources of technology transfer. This technology transfer, however, will be realized not through spin-off, but through the trickle-up trajectory.

Compatibility with Breakthroughs

The trickle-up approach illustrates the preference of Japanese managers for incremental engineering improvement over radical breakthroughs to stimulate corporate growth. A linear model of science-based innovation is often used to describe the breakthrough approach. However, almost all Japanese managers of high-tech firms view the innovation process as a spiral.

Their descriptions of the process begin with the initiating event (market opportunity or technical invention) and move sequentially through development, production, redesign, new-process technology, second-generation product, and so on.[24]

The spiral model represents a repeated sequence of linear innovations as the product is incrementally developed. This spiral innovation model can be used to illustrate Toshiba's development of the flat panel, thin-film-transistor, liquid-crystal display (TFT LCD) (see Figure 2-13).

Sometimes the spiral is represented by a two-dimensional diagram in four quadrants that represent research, development, production, and market feedback. In this type of diagram, time radially increases from the center. A product may start near the center in any of the quadrants, and spiral through all the quadrants in rapid sequence.

CONCLUSION

Although the dominant theory for diversification is based on the spin-off principle, we have found that one of the most conspicuous elements of high-tech development is the co-development of product and process technologies. This implies that diversification in high-tech industries follows a trajectory almost the opposite of that based on the spin-off principle.

We have argued that Branscomb's "trickle up" theory is not unique to the consumer electronics industry and that it can be extended to other industries, including new materials and biotechnology. Three case studies have demonstrated how growth in different industries was attained through different trickle-up trajectories.

A quantitative analysis was done, based on the assumption that one of the trajectories for downstream diversification is through the trickle-up process: a localized technical change produced by a component supplier will gradually find wider applications in its downstream product areas. The analysis revealed that the strategy is common throughout the high-tech industry and has become the dominant mode of diversification. Our regression analysis made it clear that trickle-up diversification is one method of attaining maximum business growth. This result suggests that industries outside the high-tech arena could have exploited the growth available through downstream diversification if they had followed trickle-up trajectories. A study of two Japanese companies, one in chemicals and one in textiles, provided evidence of the effectiveness of the trickle-up approach.

The management of trickle-up diversification requires a company to design a consistent business strategy that will extend its core competencies outside its conventional domain. Since companies must design business strategies to support their market, their technology, their products, and their customer relationships, four types of trickle-up diversification strategy were illustrated. Companies that organize their business around defined markets may diversify their technology. Companies that organize their business around their technology may constrain the scope of their technology to maintain leadership and diversify into many markets. Companies that organize their business around selected products may diversify by stressing their ability to switch from one technology to another. Finally, companies that organize their business around customer relationships with only a few large firms may diversify by developing the requirements for a technology with their customers.

NOTES

1. A. B. Atkinson and J. E. Stiglitz, "A New View of Technological Change," *Economic Journal* 79 (1969): 573–78.

2. L. Branscomb, "Policy for Science and Engineering in 1989: A Public Agenda for Economic Renewal," *Business in the Contemporary World* 2, no. 1 (1989).

3. Raymond Vernon, "International Investment and International Trade in the Product Cycle," *Journal of Economics* 80 (1966): 190–207.

4. F. Kodama et al., "The Innovation Spiral: A New Look at Recent Technological Advances" (paper presented at the Second U.S.-Japan Conference on High Technology and the International Environment, Kyoto, Japan, 1986).

5. Yoshikazu Ito, "A Researcher and His Motivation in Industry," in *New Roles and Societal Status of Scientists and Engineers in High Technology Era*, ed. M. Uenohara and F. Kodama (Tokyo: Mita Press, 1992).

6. F. Scherer, "Inter-Industry Technology Flows in the United States," *Research Policy* 11, no. 4 (1982): 227; K. Pavitt, "Characteristics of Innovative Activities in British Industry," *Omega* 11, no. 2 (1983): 113.

7. Prime Minister's Office of Japan, Statistical Bureau, *Report on the Survey of Research and Development* (in Japanese) (Tokyo: Nihon-Toukei-Kyoukai, 1970–1980).

8. In the case of an expense that is difficult to classify by the kind of product, the expense is divided proportionally on the basis of the number of researchers in each engineering subdiscipline.

9. S. Griliches, "R&D and Productivity: Measurement Issues and Econometric Results," *Science* 237 (1987): 31–35.

10. The values in this table are based on my calculation made for the publication: F. Kodama, "Technological Diversification of Japanese Industry," *Science* 233 (1986): 293. For this article, instead of r_i that is defined in this book, I used $r_i^\circ = [r_{i1}^\circ, r_{i2}^\circ, \ldots, r_{in}^\circ]$, in which $r_{ij}^\circ = E_{ij}/(E_{i1} + E_{i2} + \ldots + E_{in}) = E_{ij}/(\Sigma_j E_{ij})$.

Therefore, $\mathbf{r}_i^\circ$ is not a unit vector, and thus $|\mathbf{r}_i^\circ|$ does not necessarily equal 1. Instead of $\mathbf{u}_i$ and $\mathbf{d}_i$, I used $\mathbf{u}_i^\circ = [u_{1i}^\circ, u_{2i}^\circ, \ldots, u_{ni}^\circ]$ and $\mathbf{d}_i^\circ = [d_{i1}^\circ, d_{i2}^\circ, \ldots, d_{in}^\circ]$ respectively, where $u_{ij}^\circ = T_{ij}/(\Sigma_i T_{ij})$ and $d_{ij}^\circ = T_{ij}/(\Sigma_j T_{ij})$. Therefore, both upstream and downstream diversification indicators were calculated using the equations $U_i^\circ = \mathbf{r}_i^\circ \cdot \mathbf{u}_i^\circ$, and $D_i^\circ = \mathbf{r}_i^\circ \cdot \mathbf{d}_i^\circ$. I believe, however, these differences do not produce any serious biases in my relative measurements of upstream and downstream diversification. Therefore, it is not necessary to correct my findings about several causalities described in this chapter, except for those industrial sectors with vectors of $\mathbf{r}_i^\circ$, $\mathbf{u}_i^\circ$, and $\mathbf{d}_i^\circ$, which may have some particularity and/or irregularity.

11. The irregular value of rubber products may come from the combined effect of nonunit nature of the vectors $\mathbf{r}_i^\circ$, $\mathbf{u}_i^\circ$, and $\mathbf{d}_i^\circ$, used in this calculation, and the particularity of these vectors. If this is the case, the irregularity may be corrected by uses of $\mathbf{r}_i$, $\mathbf{u}_i$, and $\mathbf{d}_i$.

12. K. Clark and T. Fujimoto, *Product Development Performance: Strategy, Organization, and Management in the World Auto Industry* (Boston: Harvard Business School Press, 1991).

13. O. Grandstand, *Technology, Management and Markets* (London: Pinter, 1982), 50; C. Oskarsson, *Technology Diversification: The Phenomenon, Its Causes and Effects: A Study of Swedish Industry* (Goeteborg, Sweden: Chalmers University of Technology, 1989).

14. K. Pavitt, "Sectoral Patterns of Technical Change: Towards a Taxonomy and a Theory," *Research Policy* 13, no. 6 (1984): 343.

15. M. Dibner, "Biotechnology in Pharmaceuticals: The Japanese Challenge," *Science* 229 (1985): 1230.

16. H. Fusfeld and C. Haklish, "Cooperative R&D for Competitors," *Harvard Business Review* 85, no. 6 (1985): 60–61, 64–65, 70, 74–76.

17. E. Roberts, "New Ventures for Corporate Growth," *Harvard Business Review* 58, no. 4 (1980): 134–42; E. von Hippel, "Successful and Failing Internal Corporation Ventures: An Empirical Analysis," *Industrial Marketing Management* 6, no. 3 (1977): 163–74.

18. F. Kodama, "Policy Innovation at MITI," *Japan Echo* 6, no. 9 (1984): 66–69.

19. F. Kodama, "Direct and Indirect Channels for Transforming Scientific Knowledge into Technical Innovations," in *Transforming Scientific Ideas into Innovations: Science Policy in the United States and Japan*, ed. B. Bartocha and S. Okamura (Tokyo: Japan Society for the Promotion of Science, 1985), 198–204.

20. L. Branscomb and F. Kodama, *Japanese Innovation Strategy: Technical Support for Business Visions* (Lanham, Md.: University Press of America, 1993), 81–83.

21. C. K. Prahalad and G. Hamel, "The Core Competence of the Corporation," *Harvard Business Review* 90, no. 3 (May–June 1990): 79–90.

22. E. Roberts and M. Meyer, "Product Strategy and Corporate Success," *IEEE Management Review* 19, no. 1 (1991): 4–18.

23. M. Meyer and J. Utterback, "Core Competencies, Product Families and Sustained Business Success," Working Paper No. 65-92 (Cambridge, Mass.: MIT, 1992): Sloan Working Paper no. 3410-92.

24. Kodama et al., "The Innovation Spiral: A New Look at Recent Technical Advances."

3

R&D Competition
From Dominant-Design to Interindustry Competition

In their pioneering work of fifteen years ago, Utterback and Abernathy introduced the concept of "dominant design" and suggested that the occurrence of a dominant design may alter the character of innovation and competition in an industry.[1] Since their work, the dynamics of research and development activities have been framed around the concept of dominant design.

Utterback and Abernathy argue that when a new technology arises, there is considerable uncertainty over which of several possible variants will succeed. Different ones are tried by different firms. After a period of time and competition, one or a few of the variants come to dominate the others. That domination enforces standardization so that production or other complementary economies can be sought. Effective competition takes place on the basis of cost and scale as well as product performance.

Because Utterback and Abernathy based their dominant-design theory on detailed observation of only one industry, the automobile industry, it is possible to question just how universal the theory is. There is some skepticism about the extent to which the dominant-design theory fits the experiences of the chemical products and electronics industries.[2] Indeed, recent R&D competition among high-tech firms seems to be following a *diverging* pattern rather than the converging pattern implied by the dominant-design theory. In this chapter, we will again perform case studies and formal modeling exercises to understand why the dynamics of high-tech R&D competition can no longer be explained by the dominant-design theory.

The typical evolution of high-tech can be seen in the electronics industry. In the past forty years, this industry has experienced a series of innovations during which the vacuum tube has been supplanted by the transistor, the transistor by the integrated circuit, and the integrated circuit by very large scale integration (VLSI).[3] This pattern of rapid innovation is most apparent in the recent development of dynamic random access memory (DRAM). In this area, a new product has been introduced to the market almost every three years and before the old product has gone through the required learning curve. Furthermore, whenever a new product has appeared, it has totally replaced old products within six years.

An in-depth case study of DRAM development will reveal that the Japanese approach to technology and management in the semiconductor industry is competing with the U.S. approach. High-tech competition, then, is no longer between firms, but rather between industries. The growth of competition between industries rather than companies becomes even more apparent when we consider the development of optical fiber in Japan. In this development, solutions to basic technical problems for glass manufacturing were found in areas outside the sector. There is, then, a one-to-one correspondence between technological approach and industrial sector. Thus, the high-tech R&D process is best formulated as interindustry competition, rather than interfirm competition within a given industry.

To make informed decisions about managing competitive high-tech R&D, we need an empirical study of the R&D behavior of high-tech industries. Such a study should be quantitative and proven statistically, but to date there exists no such analysis because there is no statistical data that reflect the dynamics of R&D activity. The lack of data may be due to the difficulty in identifying an appropriate unit of analysis.

Once we have formulated the high-tech R&D dynamic as that of interindustry competition, we can develop the appropriate level of analysis. For this purpose, we will introduce the term *R&D program*, which we can define as the collection of all R&D projects carried out by a given industrial sector and focused on a certain product field. Using the database established in chapter 2, which disaggregates R&D expenditure into twenty-one sectors and thirty-two product fields, we can observe the frequency distribution of an R&D program's expenditures in each sector. Using these results, we will develop an analytical framework for R&D dynamics. If we define the state of an R&D program by its level of annual expenditure, we can build a state transition model of the R&D process in which an R&D program moves from exploratory through development states, that are governed by the R&D program's freezing-rate.

The use of different mathematical forms of freezing-rate function to represent different types of R&D competition will allow us to formulate three patterns: dominant-design, science-based, and high-tech patterns. For each pattern we can derive theoretical distribution curves concerning R&D program expenditures. By making a statistical fitting of these curves to each sector's observed frequency distribution of R&D programs, we can construct a rigorous taxonomy that will classify sectors according to the pattern they display. This taxonomy is consistent with several other studies concerning the dynamic nature of technical changes.

Once we have a dynamic characterization of high-tech competition, we can comprehend the horizontal coordination among strategic decision units found in Japanese high-tech companies.[4] Selected case studies will illustrate how Japanese companies are managing the challenges that characterize high-tech competition.

HIGH-TECH COMPETITION

A number of authors have conducted empirical studies to test the universality of the dominant-design theory.[5] In their most recent work, Utterback and Suarez begin with this hypothesis: prior to the appearance of a dominant design, there will be a wave of entering firms with many versions of the product; following the dominant design, there will be a wave of exits and consolidation in the industry.[6] Therefore, they argued, the sum of the two curves (the total number of participants in the product market segment at any point of time) will usually start with a gentle rise, followed by a much sharper rise, representing a wave of imitating firms, peak at the point when a dominant design is introduced, and then decline rather sharply until it reaches the stable condition.

In short, they tried to prove the existence of dominant design indirectly, through a time-series counting of participating companies in a given product market. Their study includes: automobiles, mechanical typewriters, television and television tubes, transistors, integrated circuits, electronic calculators, and supercomputers. Their curves for auto-

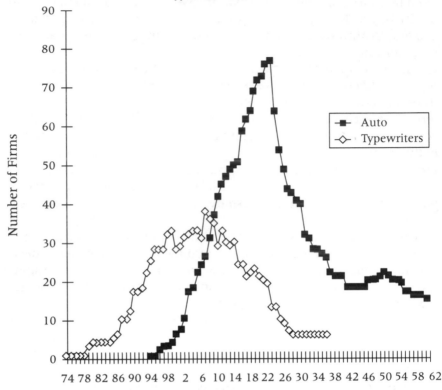

FIGURE 3-1

Summary chart of number of participants in the U.S. automobile and mechanical typewriter industries

Years (1874–1962)

Source: J. Utterback and F. Suarez, "Innovation, Competition, and Market Structure," *Research Policy* 22 (1993): 8. Reprinted by permission.

mobiles and mechanical typewriters can be seen in Figure 3-1. As shown in the figure, the total number of firms declined steadily after a dominant design was established until a point of stability was reached. Furthermore, the peak of the curve coincides with the identified date of dominant design. The year with the largest number of firms in the automobile industry, 1923, was the year that Dodge introduced the all-steel, closed body automobile, which became the dominant design. The features synthesized in the Underwood Model 5 typewriter became the dominant typewriter design in 1906.

For other industries, especially the integrated circuit industry, however, the data does not confirm their hypothesis. The curve for the

FIGURE 3-2
Number of firms participating in the U.S. integrated circuits industry

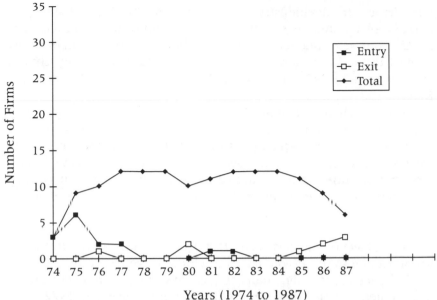

Years (1974 to 1987)

Source: J. Utterback and F. Suarez, "Innovation, Competition, and Market Structure," *Research Policy* 22 (1993): 14. Reprinted by permission.

integrated circuit industry shows no clear peak, and there is a very broad plateau in total industry participation that is marked by continuing entries and exits (see Figure 3-2). Having obtained these unexpected results, Utterback and Suarez suggest that no one product of any generation of integrated circuits can be considered a dominant design. They assert:

> The integrated circuit has kept on changing substantially from generation to generation. In DRAM, competition has been tough both within and between generations. No one firm has been able to maintain a leadership from one generation to another. In general, American firms have been losing ground to Japanese entrants.[7]

We can generalize this observation by stating that the concept of dominant design appears less applicable and effective in the high-tech industry. Therefore, the high-tech industry is not following the converging pattern implied by the dominant-design theory. Instead, competition among high-tech firms appears to be following a diverging pattern.

Diverging Pattern

The microelectronics industry offers the most discernible diverging pattern of research and development activities. Innovations in MOS-type (metal oxide semiconductors) DRAM, led to the introduction of a new product to market almost every three years: 1 Kbits memory chip in 1972, 4 Kbits in 1975, 16 Kbits in 1978, 64 Kbits in 1981, and 256 Kbits in 1983.

In high-tech competition, the price of the product decreases over time, and the extent of production increases, but when a new product is introduced to the market, production of the preceding product decreases drastically even though its price continues to decrease. Therefore, manufacturers cannot survive by selling the old products; they must invest in R&D to produce a new generation of products or go out of business. Indeed, a glimpse at the evolution of DRAM in the past two decades demonstrates just how remorseless innovation at the leading edge of technology has been.[8] It shows, too, how destructive innovation can be to those who fail to catch the wave (see Table 3-1).

The first commercial memory chip could store *1,000 bits* of computer data. When the so-called 1k DRAM was introduced in 1972, only U.S. manufacturers knew how to make it. No sooner had ten U.S. semiconductor firms begun to earn profits from it than its replacement, the 4k chip, hit the market. As shown in Table 3-1, three manufacturers (General Electric, Corning, and Westinghouse) missed the wave and were effectively forced out of the business. However, three new manufacturers (Intel, Rockwell, and Signetics) plunged into the game.

By 1978, the 16k DRAM had become the standard memory chip. Again, three American manufacturers (Rockwell, General Instruments, and American Micro Devices) had failed to catch the wave. Suddenly, three Japanese manufacturers (NEC, Hitachi, and Toshiba) were among the world's top ten memory chip makers. The Japanese then took the lead. The 64K chip was launched in 1981, a year earlier than expected. By the time the 256k chip arrived in 1984, again a year earlier than expected, two more Japanese manufacturers had entered the market. This time practically all the manufacturers were Japanese.

Although we can see that some companies suddenly appeared and disappeared, the greater change has been the shift in world leadership from the United States to Japan. We can hypothesize that the shift has occurred because of the different approaches to technology and management taken by the semiconductor industries in these two countries. At the microlevel, there appear to be discernible differences in American and Japanese approaches and attitudes.

TABLE 3-1
Changes in the world's top ten chip makers

Ranking	1972 1k DRAM	1975 4k	1978 16k	1981 64k	1984 256k	1987 1M	1990 4M
1	TI	TI	TI	Motorola	*Hitachi*	*Toshiba*	*Hitachi*
2	Motorola	Fairchild	Motorola	TI	*NEC*	*Hitachi*	*Toshiba*
3	Fairchild	National	National	*NEC*	*Fujitsu*	*Mitsubishi*	*NEC*
4	RCA	Intel	Intel	*Hitachi*	*Toshiba*	*NEC*	*Fujitsu*
5	GE	Motorola	*NEC*	National	AT&T Tech	*Oki*	*Mitsubishi*
6	National	Rockwell	Fairchild	*Toshiba*	*Mitsubishi*	*Fujitsu*	*Samsung*
7	G.I.	G.I.	*Hitachi*	Intel	*Oki*	TI	*Oki*
8	Corning	RCA	Signetics	Phillips	TCMC	*Matsushita*	—
9	W.H.	Signetics	Mostek	*Fujitsu*	TI	—	—
10	AMD	AMD	*Toshiba*	Fairchild	Intel	—	—

Sources: For 1972–1981 data, see D. Okimoto, T. Sugano, and F. Weinstein, eds., *Competitive Edge: The Semiconductor Industry in the U.S. and Japan* (Stanford: Stanford University Press, 1984); for 1984–1987 data, see V. Nicholas, "Thinking Ahead: A Survey of Japanese Technology," *The Economist* 313, no. 7631 (1989): 2–8; for 1990 data, Dataquest Inc.

Note: Japanese companies are in *italics*.

A study made by D. Okimoto and Y. Nishi asserted that these differences have become apparent at various points in the history of U.S.-Japanese semiconductor competition.[9] In designing the 256k DRAM, several U.S. companies abruptly discarded the old planar processes, believing that planar technology was bumping up against physical limits and that those limits would be reached in the 1M DRAM. Thus, American managers were anxious to develop a new trench technology as early as possible in order to dominate future ultra large scale integration (ULSI) markets. United States companies found to their dismay, however, that developing trench technology took more time than they anticipated. In contrast, Japanese semiconductor companies continued with the more familiar, less risky planar technology and sought to push this technology to its outer limits while doing research on trench technology. As a result, Japanese companies were able to dislodge their U.S. competitors from their commanding positions as the world's largest manufacturers.

The planar-trench situation seems to be repeating itself in the choice between phase-shift technology and X-ray lithography. United States companies, seeing a dead-end ahead for the current generation of etching technology, have invested heavily in X-ray lithography for submicron circuits in future generation commodity chips. IBM and other U.S. companies have now all but abandoned phase-shift technology and concentrated the bulk of their resources on X-ray lithography, a technology still far from commercial application. Meanwhile, all of the major Japanese companies have picked up phase-shift technology and appear ready to utilize it commercially, starting with the 64M DRAM, now in the prototype phase of development. At the same time, Japanese firms are working on X-ray lithography in corporate labs as well as in a national research consortium.

The differences in management approaches are most visible when we consider U.S. and Japanese decisions to invest in R&D projects. For Japanese managers, decisions to invest in certain R&D projects are based on the best way to meet the technological requisites of global competitiveness. In contrast, the overriding concern of U.S. firms is to maximize financial returns on investment. To this end, U.S. managers have adopted a variety of quantitative methodologies, including discounted cash flow, to assess risk and return. Although Japanese companies also subject their investments to financial analysis, their analytical tools are not as sophisticated as discounted cash flow. Instead, managers rely on in-depth discussions of the pros and cons of moving in certain technological directions, paying careful attention to the explicit and hidden assumptions underlying scenarios of likely outcomes.

Rather than allowing financial considerations to dictate R&D decisions, Japanese managers place their emphasis on what it will take technologically to be world competitors and how to make the best use of their research talent to reach the company's R&D goals.

Interindustry Competition

The industry-specific nature of high-tech competition becomes even more obvious when we consider the development of optical fiber. In the United States Corning Glass Works (CGW), a glass manufacturer, has been the leader throughout the development of optical fiber. In Japan, however, leadership in the development of optical fiber has switched from industry to industry, from glass manufacturers to communication device manufacturers and from communication enterprises to cable manufacturers, while joint research has spanned the boundaries of all these industries. Indeed, the most dramatic innovations have occurred when the leaders emerged from a new sector.

The fiber optics developed by the Japanese glass industry lacked mechanical strength and caused transmission loss. Cable manufacturers solved the problem of mechanical fragility by developing a coating technology. The problem of transmission loss was resolved by Nippon Telegraph and Telephone Corporation (NTT), a major user of optical fibers, which discovered that transmission loss could be greatly reduced by using a wave length much longer than the one usually used. In 1969, the joint staff of Nippon Sheet Glass (NSG) and Nippon Electric Company (NEC) produced a gallium-doped fiber that had transmission loss of 80 dB/km. In the United States, CGW produced a TiO_2-doped fiber that had a transmission loss of 20 dB/km, earning the company worldwide acclaim. It should be noted that the transmission loss of 20 dB/km is comparable to that of coaxial cables. Glass manufacturers such as NSG and CGW, therefore, were foremost among those companies that made great contributions during the period when the potential of optical fiber was being researched. In the years that followed, an optical fiber with a transmission loss equivalent to that of guided millimeter wave transmission lines was developed.

At this point, development was directed at verifying the usefulness of optical fibers in place of coaxial cables and guided millimeter wave transmission systems. The leaders in innovation at this stage were the research laboratories of telephone companies. However, before optical fibers could be used in actual working environments, their mechanical strength had to be verified. Cable manufacturers that had previously produced copper wires for communications media played a leading role in this research by developing a coating technology that worked. Tandem Double Coating was developed by SEI in 1974. Thus, during the verification period leadership switched from glass manufacturers to cable manufacturers and telecommunications companies.

Once the point of practical application was reached, the problem centered around use and system reliability. It would be risky to attempt to use the system based on the new technology alone. Therefore, the leading players became the telephone operating companies that would benefit from the system. In 1975 NTT and SEI formed a joint research team. It found that reductions in transmission loss could be better attained by using a wavelength of 1.3 μm, for example, which was longer than the wavelength of 0.85 μm achieved with GaAs semiconductor lasers. The discovery extended repeater-to-repeater distance and eliminated the need for high-power lasers, thus making fiber optics competitive. At the same time, it opened the way to the development of new semiconductor lasers, such as the InP-type, which can operate in long wavelengths at normal temperatures. Thus, optical communications proved to be better suited for long-distance communication than any other available system. In 1979, NTT confirmed a transmission loss of as low as 0.2 dB/km at 1.55 μm, thus establishing a solid foundation for practical application.

With public recognition of the capabilities and reliability of the fiber-optic system, demand rose substantially. This stage required technology for mass production. The need was first met by Bell Telephone Laboratories (BTL) and CGW in the United States, which developed modified chemical vapor deposition (MCVD). Then, NTT and SEI provided the vapor phase axial deposition (VAD) method. In other words, the principal players in Japan at this stage were the research organs of the would-be largest users of the system and the companies promoting fiber-optics business.

We can use this case study to create an innovation model. A cyclical process of research, development, production, and distribution is called an *innovation cycle*.[10] Our review of the history of optical fiber development in Japan revealed that the leading innovators at each stage came from different industrial sectors while collaborating in joint research across industry boundaries. Furthermore, each time a change in leaders occurred, dramatic improvements in technological development were made. Therefore, the innovation cycle should be thought of as multilayered, as illustrated in Figure 3-3, because high-tech R&D is carried out simultaneously in a wide variety of industries.

In this multilayered structure, changes in leaders could be taken to mean that leaders move from an innovation cycle in one industry to

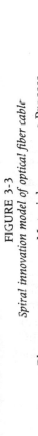

FIGURE 3-3

Spiral innovation model of optical fiber cable

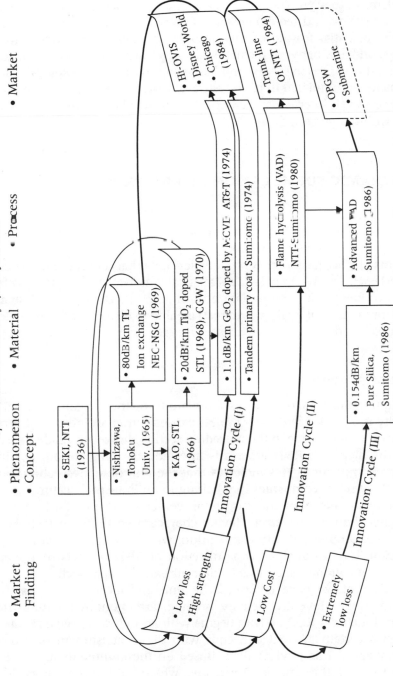

- Market Finding
- Phenomenon
- Concept
- Material
- Process
- Market

Source: Tsuneo Nakahara, SEI (paper presented at the Seminar on the Development of High Technologies in the U.S.A. and Japan: Comparative Evaluation and Policy Implications, John F. Kennedy School of Government, Harvard University, 1992). Reprinted by permission.

109

a new innovation cycle in a different industry, solving technological problems as they go. The innovation cycle, then, becomes a *spiral innovation model* with three-dimensional cycles.[11]

The essential feature of this innovation model is the one-to-one correspondence between technological approach and industrial sector. Each industry tries to solve a problem using specific technological competencies accumulated in its industrial sector. Therefore, the high-tech R&D process is interindustry competition, instead of interfirm competition within a given industry.

A DYNAMIC FORMULATION OF R&D PROGRAMS

To date there has been no quantitative, statistically proven analysis of R&D dynamics, although there are some conceptual models that are neither quantitative nor statistical.[12] One of the major reasons for this lack may be because there is no statistical data that reflect the dynamics of R&D activity. We may lack a statistical data because we have so far failed to identify an appropriate unit of analysis. Once we choose an appropriate unit of analysis, we may be able to discover a database for empirical studies.

Unit of Analysis

The minimum unit of analysis for a study of the dynamics of R&D activity is an individual R&D project carried out at an individual firm. The case studies made in the preceding section, however, made it clear that the high-tech R&D process is interindustry competition, not interfirm competition. This indicates that the unit of analysis should be the whole set of companies that belong to an industrial sector.

In terms of level of analysis, then, we need to think of a level that is higher than project level. Having characterized the high-tech R&D process as an interindustry competition, we can conclude that the appropriate level of analysis is a collection of R&D projects that share something common such as a similar aim and/or approach.

Now that we have described the unit and level of analysis, we should think of inventive activity as being carried out by the industrial sector. Industrial R&D activity begins with an industrial sector's interests in a certain product and ends with the establishment of a new technology in this product field. Based on the arguments above, we can introduce the term *R&D program*, which we will define as "the collection of all R&D projects carried out by a given industrial sector and focused on a certain product field."

Having determined the unit of analysis, we can now find a database for an R&D program. Since our unit of analysis is the R&D activities of an industrial sector that are focused on a certain product field, the disaggregation of R&D activities by industrial sector and by product fields is a necessity. Once again, the R&D data found in the *Survey of Research and Development* can provide us with a database for an R&D program, as the survey disaggregates a company's intramural R&D expenditure into different product fields.[13]

We must again distinguish between a sector's expenditures within principal product fields and its expenditures outside these product fields. Compared to R&D expenditures within principal product fields, a sector's R&D expenditures outside its principal product fields can be assumed to reflect the dynamics of an R&D program because an increase in expenditure can only be expected if there is success in earlier stages. If the prospects are not favorable, these expenditures can be frozen because they are investments outside the main business activity.

In the case of the expenditures within principal product fields, however, we cannot assume an increase is an indicator of success.[14] Thus, a sector's R&D expenditure within its principal product fields does not necessarily reflect the dynamics of an R&D program. For this reason we will confine our database to a sector's R&D expenditures outside its principal product fields.

Before beginning our analysis, therefore, we have to distinguish principal product fields and fields that are not principal product fields for each sector. By following the procedure adopted in chapter 2, we will make this distinction but only among manufacturing industries. This time, we will use a more detailed classification. Manufacturing industries are disaggregated into twenty-one industrial categories, and the twenty-five product categories produced by these manufacturing industries are assigned to any one of the industrial sectors (see Table 3-2). With this classification, we have aggregated twenty-five product fields into the twenty-one groups.

Frequency Analysis

Now we can test our assumption that a given sector's R&D expenditure outside its principal product fields reflects the dynamics of an R&D program. For this test, we can use the frequency distribution of a sector's R&D expenditures into all product fields outside its principal product fields. Since our database is available for every year since 1970, we

TABLE 3-2
Classification of product fields and industrial sectors

Industrial Sector	Product Field
Food	Food
Textile mill	Textile
Pulp and paper	Pulp and paper
Printing and publishing	Printing and publishing
Industrial chemicals	Chemical fertilizers, and organic and inorganic chemicals
	Chemical fibers
Oil and paints	Oil and paints
Other chemicals	Other chemicals
Drugs and medicines	Drugs and medicines
Petroleum and coal	Petroleum
Rubber	Rubber
Ceramics	Ceramics
Iron and steel	Iron and steel
Nonferrous metals	Nonferrous metals
Fabricated metal	Fabricated metal
Ordinary machinery	General machinery
Electrical machinery	Household electrical appliances and other equipment
Communications and electronics	Communications and electronics equipment
Motor vehicles	Automobiles
Other transportation equipment	Ships
	Aircraft
	Rolling stock
	Other transportation equipment
Precision equipment	Precision instruments
Miscellaneous manufacturing	Miscellaneous

can *pool* enough data for each sector's R&D program to permit a statistical analysis.

Given E_{ijt} (*i-th* industry's R&D expense into *j-th* industry's principal product fields in *t-th* year, $i, j = 1, \ldots, N; t = 1, \ldots, T$) and D_t (research expenditure deflator of *t-th* year, the reference year is 1975), then, the *real R&D expenditure* for various product fields can be represented by $R_{ijt} = E_{ijt}/D_t$. Now, we can use each sector's expenditures in all the product fields outside its principal ones: for the *i-th* industrial sector, for example, all the observations of R_{ijt}'s ($j = 1, \ldots, N; j \neq i$; $t = 1, \ldots, T$) are available. Therefore, we can pool $(N - 1) \cdot T$ sample points for each industrial sector.

Using these pooled sample-points, we can draw the frequency dis-

FIGURE 3-4
*Selected examples of frequency distribution of R&D expenses outside
principal product field*

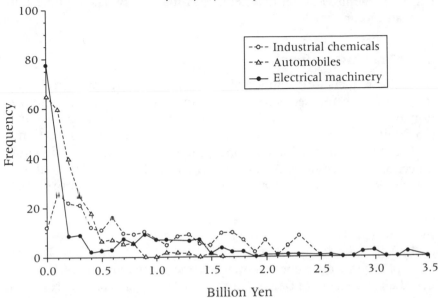

Billion Yen

tribution curve of each sector's R&D program. In order to illustrate this type of frequency analysis, we selected three industrial sectors: industrial chemicals, electrical machinery, and motor vehicles. For these sectors, the frequency distribution of each sector's expenses into all the product fields outside its principal ones, is shown in Figure 3-4. Because we used the time-series data for the period 1970–1982, we can use 260(= 20 · 13) sample-points for each sector. As we can see, all the three distribution curves appear to follow exponential distribution. This exponential distribution implies that most of an industry's R&D programs are for exploratory search. These programs have small budgets. A few are for development and have larger budgets. We can say, therefore, that the distribution in Figure 3-4 faithfully reflects the dynamic process of an R&D program in which a heavy investment occurs only after a great deal of exploratory research has been conducted for a long period of time and the prospects look promising. Indeed, the distribution illustrates dynamics of an R&D program— almost all of the exploratory research programs turn out to be failures and few of them survive for development.

Although the three distribution curves appear to follow the exponential distribution, a closer look reveals a difference among the industrial sectors. While the curve for industrial chemicals is that of smoothly decreasing function, curves for electrical machinery and motor vehicles decrease sharply. Furthermore, electrical machinery has a longer tail-off than motor vehicles.

State Transition Model

Using the frequency distribution of annual expenditures, we can characterize each industrial sector's R&D program, but it is a static characterization. However, let us assume that this static frequency distribution curve is a result of a dynamic process, which we will describe in this section.

First of all, we need a key concept that will describe the dynamic aspect of R&D activities. Although it is widely held that innovation is a dynamic process, less attention is given to the dynamic nature of R&D activity.[15] The well-established classification scheme for R&D activity divides the activity into *basic* research, *applied* research, and *development* research. This is a static, not dynamic, characterization.

Dissatisfied with the conventional classification, the U.S. Department of Defense developed its own scheme. To manage its R&D activities, the Department of Defense divides R&D into *fundamental* research, *exploratory* development, *advanced* development, *engineering* development, and *operational systems* development.[16] Because of the skewed cost distribution among various stages of R&D activity, cost becomes a very important factor. The Department of Defense found that the total cost of R&D is distributed as follows: 5 percent is allocated to fundamental research, 10 percent to exploratory development, 20 percent to advanced development, 50 percent to engineering development, and 15 percent to operational systems development.

What is important here is the distinction made between an *exploratory state* and *development state,* in which the first two stages given above (fundamental research and exploratory development) are categorized as the exploratory state, and the remaining three stages (from advanced development through operational systems development) are categorized as the development state. With this distinction, we can think of a dynamic process in which an R&D program shifts from an exploratory state to a development state. Through this dynamic process, an exponential type of frequency distribution of an R&D program's expenditure will be generated. Based on this interpretation of the exponential frequency distribution, we can describe a *state transition* model of an R&D program as follows: an R&D program stays in a continuous transition process from the exploratory state to the development state as long as its prospects are favorable. However, at any stage in the process when its prospects are found to be unfavorable, transition is frozen.

Let us formulate this process of state transition more rigorously. To do so, we have to define the *state* of an R&D program. It can be defined by the level of annual expenditure for an R&D program. It is denoted by C. The exploratory state can be defined as that level where expenditure C is smaller, and the development state can be defined as that level where C is larger. Now we can introduce a conceptual device that governs an R&D program's transition from the exploratory state to the development state. Our device is the *freezing rate* of an R&D program, which we can define as "the probability that an R&D program's annual expenditure is frozen at C." Since this probability is dependent upon an R&D program's state defined by C, it is called the *freezing-rate function* and denoted by r(C).

By drawing a frequency distribution curve, we have obtained the probability that an R&D program's annual expenditure is C. This curve represents the probability-density function of C and is denoted by f(C). We can derive its probability function, R(C), i.e., the probability that the annual expenditure level of an R&D program is above C, by using an obvious relationship between f(C) and R(C):

$$f(C) = d/dC[1 - R(C)] = -R'(C). \qquad (1)$$

Now let us say that the dynamic process in which state transition is governed by the freezing rate will produce the observed frequency distribution curve, which represents a static characterization of an R&D program. Having made this interpretation of the probability-density function, the freezing-rate function, r(C), can be formulated as the *conditional probability* that an R&D program's annual expenditure is frozen at C, given that it has come to the state of C. Therefore, r(C) can be represented by

$$r(C) = f(C)/R(C) \qquad (2)$$

By replacing f(C) with the relationship depicted in equation (1), r(C) can be further represented by

$$r(C) = -R'(C)/R(C). \qquad (3)$$

This equation implies that all the differences in frequency distribution curves among industrial sectors can now be deducted from the difference in the freezing-rate function. In other words, once we specify the mathematical form for the freezing-rate function, $r(C)$, we can derive the frequency distribution, $f(C)$. Based on the relationship depicted in equation (3), we can derive $R(C)$ by integration,

$$R(C) = \exp\left[-\int_{\phi=0}^{\phi=C} r(\phi) \cdot d\phi \right]. \tag{4}$$

By replacing $R(C)$ with equation (4) in the equation (2), we can derive the formula for transforming $r(C)$ into $f(C)$ as follows:

$$f(C) = r(C) \cdot \exp\left[-\int_{\phi=0}^{\phi=C} r(\phi) \cdot d\phi \right]. \tag{5}$$

Once we have done this, we can formulate the converging pattern implied by the dominant-design theory by assigning a specific mathematical form to the freezing-rate function. We can assume that the freezing-rate function $r(C)$ is a decreasing function of C. At the exploratory state, defined as the state in which expenditure C is smaller, the rate at which the annual expenditure level is frozen can be assumed to be higher. At the development state, defined as the state in which C is larger, the freezing rate can be assumed to be lower because the prospects of an R&D program in the development state have already been proven during the exploratory state that precedes it.

The converging pattern implied by the dominant-design theory suggests that there is no freezing once a program enters the development state. Therefore, we can assume that the freezing rate approaches zero in that state. From the possibilities available to us, we have chosen the following function:

$$r(C) = b \cdot \exp(-d \cdot C). \tag{6}$$

Because this is a mathematical representation of the dominant-design theory, I call it the *dominant-design pattern*. Using the transformation formula of equation (5), we can derive the probability density function for this freezing-rate function. For the pattern represented by equation (6), it follows that

$$\int_0^C r(\phi) \cdot d\phi = -(b/d) \cdot [\exp(-d \cdot C) - 1]. \tag{7}$$

By substituting equations (6) and (7) for two terms in the transformation formula of equation (5), we can derive $f(C)$ as follows:

$$f(C) = b \cdot \exp[(b/d \cdot \{\exp(-d \cdot C) - 1\} - d \cdot C]\tag{8}$$

Statistical Test for Dominant-Design

Using the curve fitting between the observed and the theoretical values, we can try to determine statistically whether or not the dominant-design pattern is plausible for each industrial sector. To do so, for each industrial sector's observed distribution curve, as illustrated in Figure 3-4, we will try to fit the theoretical probability-density function as derived in equation (8). Because of the complicated mathematical functional form of the derived probability density function, we must employ a nonlinear, least square method, i.e., the Marquard method. This is one of the steepest descent iteration methods for nonlinear curve fitting. Although the theoretical basis for significance tests of the parameters of nonlinear functions is not well established, we followed the same procedures as for linear functions.

The probability density function of the dominant-design pattern is derived by assuming the freezing-rate function as: $r(C) = b \cdot \exp(-d \cdot C)$. Therefore, the derived probability density function has two parameters, b and d. The results of the curve fitting are shown in Table 3-3. This table includes the coefficient of determination in the first column as a measure of the degree of fitness and the estimated values of the two parameters, together with their t-values in parentheses as a measure of significance level, in the fourth and sixth columns. We could not complete the estimation process in the miscellaneous manufacturing sector because convergence conditions were not met; i.e., the observed distribution curve was so irregular that the unique values for the two parameters could not be determined. This may be inevitable because the category includes so many different kinds of industry.

In the twenty other industrial sectors, we obtained a fairly high degree of fitness (the coefficients of determination are high). Therefore, we might argue that the dominant-design theory more or less holds for all the sectors. Because the coefficients of determination on average are high, the level of significance (the t-value) is a critical determinant of the dominant-design pattern. By following the same procedure developed for linear functions, we can discover whether or not the t-values for the two parameters are larger than 2.02, the significance level of 95 percent. If they are, we cannot deny that an industrial sector is following the dominant-design pattern. In only one sector, industrial

TABLE 3-3

Statistical test of the dominant-design pattern

Industrial Sector	Coefficient of Determination	b	(t-value)	d	(t-value)
Food	0.9142	0.44	(32.15)	0.40	(8.48)
Textile	0.9197	0.33	(26.36)	0.19	(5.67)
Pulp and paper	0.9965	0.57	(65.08)	0.57	(16.14)
Printing/publishing	0.9204	0.53	(12.92)	0.32	(2.45)
Industrial chemicals	*0.8864*	*0.08*	*(33.68)*	*0.01*	*(1.85)*
Oil and paints	0.8967	0.45	(26.68)	0.32	(6.13)
Drugs and medicines	0.9889	0.54	(66.34)	0.68	(18.01)
Other chemicals	0.9683	0.38	(43.69)	0.26	(10.37)
Petroleum and coal	0.9722	0.48	(37.51)	0.52	(10.42)
Rubber	0.9153	0.45	(15.19)	0.27	(3.03)
Ceramics	0.9111	0.30	(28.06)	0.15	(5.39)
Iron and steel	0.9451	0.30	(35.50)	0.23	(9.91)
Nonferrous metals	0.9388	0.28	(24.36)	0.14	(4.80)
Fabricated metal	0.9571	0.30	(30.06)	0.19	(7.25)
Ordinary machinery	0.7894	0.18	(19.36)	0.14	(6.36)
Electrical machinery	0.9344	0.30	(100.40)	0.63	(37.09)
Comm. and electronics	0.9885	0.45	(273.30)	0.52	(79.87)
Motor vehicles	0.9643	0.28	(53.76)	0.15	(11.06)
Other transportation	0.9755	0.40	(112.10)	0.27	(25.82)
Precision equipment	0.9813	0.42	(52.28)	0.47	(15.58)
Miscellaneous manufacturing	—	—		—	

chemicals, do we find that the t-value for *d* is below 2.02. Therefore, the industrial chemicals sector does not follow the converging pattern derived from dominant-design theory.

IDENTIFYING OTHER PATTERNS

If the R&D program's behavior in industrial chemicals does not follow the dominant-design pattern, what other possible patterns can we think of? First of all, we can think of a pattern in which there is no discernible regularity at all. We can assume that an R&D program's freezing rate is independent of its state as defined by its annual expenditure level. If we assume this, we assume that an R&D program's freezing process follows a random process in transition from the exploratory to the development state. This pattern is observed in industries in which scientific findings are directly related to the main business activity. Therefore, we call it the *science-based pattern*.

Science-Based Pattern

In this pattern, freezing-rate function $r(C)$ is independent of C. Therefore, the mathematical function for the science-based pattern is

$$r(C) = a.$$

We can derive an exponential distribution curve as the probability density function $f(C)$ as follows:

$$f(C) = a \cdot \exp(-a \cdot C). \qquad (9)$$

With this, we are ready to make a curve fitting of the science-based pattern for all twenty-one industrial sectors, including miscellaneous manufacturing. Since the probability density function of the science-based pattern is derived by assuming the freezing-rate function to be $r(C) = a$, the derived probability density function has only one parameter. The result of the curve fitting of the science-based pattern is shown in Table 3-4. Because we can see that the t-value of the industrial

TABLE 3-4
Statistical test of the science-based pattern

Industrial Sector	Coefficient of Determination	a	t-value for a
Food	0.8532	0.45	(26.64)
Textile	0.8991	0.33	(25.71)
Pulp and paper	0.9264	0.56	(15.14)
Printing and publishing	0.9057	0.54	(13.00)
Industrial chemicals	*0.8858*	*0.08*	*(37.77)*
Oil and paints	0.8684	0.46	(25.77)
Drugs and medicines	0.8782	0.53	(21.05)
Other chemicals	0.9293	0.38	(31.61)
Petroleum and coal	0.8786	0.47	(19.09)
Rubber products	0.8966	0.46	(15.02)
Ceramics	0.8953	0.30	(27.99)
Iron and steel	0.8865	0.29	(27.04)
Nonferrous metals	0.9193	0.27	(23.08)
Fabricated metal	0.9213	0.29	(24.06)
Ordinary machinery	0.7169	0.15	(18.20)
Electrical machinery	0.6558	0.23	(42.30)
Communications and electronics	0.8776	0.42	(89.97)
Motor vehicles	0.9482	0.28	(49.00)
Other transportation	0.9452	0.40	(80.57)
Precision equipment	0.8738	0.40	(21.31)
Miscellaneous manufacturing	0.8631	0.17	(18.90)

chemicals sector is now 37.77, much larger than 2.02, the significance level of 95 percent, we can say the industrial chemicals sector follows the science-based pattern.

The High-Tech Pattern

In the freezing-rate function shown in equation (6), which was formulated for the dominant-design pattern, it was assumed that there is no freezing once a program enters the development state. The dominant-design pattern, however, assumes interfirm, not interindustry, competition, and our case studies have shown that R&D competition in the high-tech industry can be better formulated as interindustry competition. Interindustry competition suggests that an R&D program can be frozen even when the program enters the development state. If someone else finds a technology that can match the performance level of products based on a different principle, programs in the development state will be frozen. Therefore, we need to choose another function in which the freezing rate approaches a positive value. We can call this the *high-tech pattern*. The three freezing-rate functions are illustrated and compared in Figure 3-5.

From the possibilities, we have chosen the following function:

$$r(C) = f \cdot \exp(-g \cdot C) + h, \qquad (h > 0). \qquad (10)$$

By following the same derivation procedure used for the dominant-design pattern, we can derive the probability density function $f(C)$ for the high-tech pattern as follows:

$$
\begin{aligned}
f(C) = &[f \cdot \exp(-g \cdot C) + h] \\
&\cdot \exp[(f/g) \cdot \{\exp(-g \cdot C) - 1\} - h \cdot C]
\end{aligned}
\qquad (11)
$$

The derived probability density function is highly nonlinear and fairly complicated. Therefore, we have tried to visualize the shape of this distribution curve as compared with those of the other two freezing-rate functions by assigning arbitrary values to parameters of the function (see Figure 3-6). As we can see, the curve for the science-based pattern follows the exponential distribution. The curves for the dominant-design and high-tech patterns, however, decrease more sharply than the exponential curve does. Furthermore a long plateau is visible after an R&D program's expenditure exceeds a certain level in the high-tech pattern while it is not visible in the dominant-

FIGURE 3-5
Three types of freezing-rate function

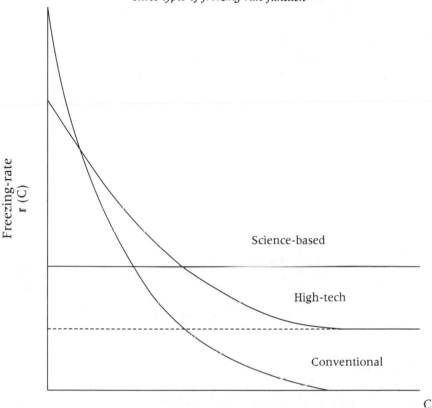

An R&D program's annual expenditure

design pattern. This difference is the same difference we observed between the electrical machinery sector and the motor vehicle sector in Figure 3-4.

The probability density function of the high-tech pattern is derived by assuming the freezing-rate function to be $r(C) = f \cdot \exp(-g \cdot C) + h$. Therefore, the derived probability density function has three parameters, f, g, and h. The results of fitting the high-tech pattern are shown in Table 3-5. If we assume that $h = 0$, however, we fail to discriminate the high-tech pattern from the dominant-design pattern. Therefore, this table contains only the t-values for the estimated parameter values of h. As we see, the t-values for h are lower than 2.02 except in five sectors (drugs and medicines, ordinary machinery, electrical machinery, communications and electronics, and precision equip-

FIGURE 3-6
Differences in three probability density functions

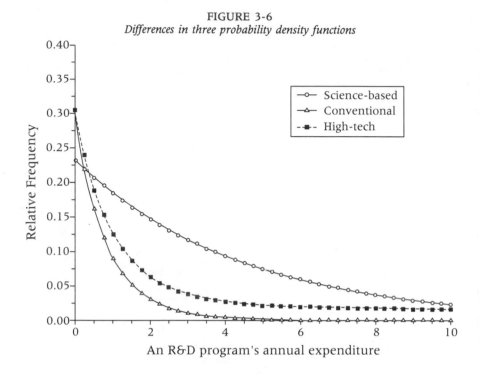

An R&D program's annual expenditure

ment). Therefore, except for these five sectors, we cannot reject the null hypothesis that $h = 0$. If h equals zero, the freezing-rate function defined by equation (10) for the high-tech pattern becomes equivalent to that defined by equation (6) for the dominant-design pattern.

In all five sectors, the coefficient of determination is better than it is in either of the other patterns. For example, in drugs and medicine, the coefficient of determination has moved from 0.9889 in the dominant-design pattern to 0.9900, in ordinary machinery from 0.7894 to 0.8013 and in electrical machinery from 0.9344 to 0.9569. Combining these improvements with our knowledge of the t-values, we can conclude that the high-tech pattern is apparent only in drugs and medicines, ordinary machinery, electrical machinery, communications and electronics, and precision equipment.

Using Tables 3-4, 3-5, and 3-6, we can now identify the sectoral patterns for twenty of the twenty-one industrial sectors. To do so, we must choose the best freezing-rate function for each sector from the three possible candidates. The criteria for the choice are the degree of fitness and the level of significance. In other words, we look for a high

TABLE 3-5
Statistical test of the high-tech pattern

Industrial Sector	Coefficient of Determination	h	t-value for h
Food	0.9145	0.0029	(0.52)
Textile	0.9198	0.0015	(0.14)
Pulp and paper	0.9968	0.0083	(1.13)
Printing and publishing	0.9212	−0.0229	(0.36)
Industrial chemicals	0.8865	−0.0274	(−0.08)
Oil and paints	0.8967	0.0009	(0.11)
Drugs and medicines	*0.9900*	*0.0077*	*(2.23)*
Other chemicals	0.9683	0.0022	(0.40)
Petroleum and coal	0.9735	0.0095	(1.43)
Rubber products	0.9154	−0.0069	(−0.19)
Ceramics	0.9111	0.0004	(0.04)
Iron and steel	0.9181	0.0065	(1.10)
Nonferrous metals	0.9388	−0.0005	(−0.03)
Fabricated metal	0.9375	0.0047	(0.48)
Ordinary machinery	*0.8013*	*0.0356*	*(2.97)*
Electrical machinery	*0.9569*	*0.0294*	*(21.19)*
Communications and electronics	*0.9886*	*0.0008*	*(3.32)*
Motor vehicles	0.9644	0.0016	(0.39)
Other transportation	0.9755	0.0002	(0.22)
Precision equipment	*0.9841*	*0.0118*	*(2.75)*

coefficient of determination and t-values for all the parameters that are larger than the values specified by a significance level. The results of our identification are shown in Table 3-6. As seen in the table, only one industry, industrial chemicals, follows the science-based pattern. Five industries, drugs and medicines, ordinary machinery, electrical machinery, communications and electronics, and precision equipment follow the high-tech pattern. Fourteen industries, from food through other transportation equipment, follow the dominant-design pattern. Finally, we cannot identify miscellaneous manufacturing because the estimation process is not completed in the curve fitting of the two function types.

Other Empirical Evidence

Although we have established an empirical taxonomy of R&D activities, there are still two tasks to be accomplished: one concerns the definitions of categories in our taxonomy, and the other concerns the

TABLE 3-6
Results of sectoral identification

Typology	Industrial Sector
Dominant-design:	Food; Textile; Pulp and paper; Printing and publishing; Oil and paints; Other chemicals; Petroleum and coal; Rubber; Ceramics; Iron and steel; Nonferrous metals; Fabricated metal; Motor vehicles; Other transportation equipment
Science-based:	Industrial chemicals
High-tech pattern:	Drugs and medicines; Ordinary machinery; Electrical machinery; Communications and electronics; Precision equipment
Unidentified	Miscellaneous mfg.

taxonomy's effectiveness and usefulness. In our model formulation, we defined the dominant-design pattern as one in which an R&D program is in no danger of being frozen after it has reached the development state, i.e., there is no risk once a key technology is established. A "science-based pattern" was defined as one in which the danger of an R&D program being frozen remains at the same level in all states. Finally, the "high-tech pattern" was defined as one in which the danger of being frozen remains at a certain level even after an R&D program reaches the development state.

Our taxonomy, however, is based only on the differences in the way in which risks are involved throughout an R&D program. Furthermore, each category was named independently of our model formulation. Therefore, we have to fill the discrepancy between how each pattern is named and how it is defined. To test the effectiveness of our taxonomy, we will compare its explanatory power to the explanatory power of other theories on the dynamic nature of technical changes put forth by academic and policy-making communities.

Science Linkage. It is often said that industrial chemicals, pharmaceuticals, and electronics industries are similar to each other because of their reliance on science. In his taxonomy of technical change, for example, K. Pavitt classified electronics/electrical equipment and chemicals as science-based.[17] Because his taxonomy does not describe pharmaceuticals separately, we can assume that he included pharmaceuticals in chemicals. However, a taxonomy with these three industries in the same category does not explain why Japan has a strong electronics industry but not a strong chemical industry, how biotechnology has changed the pharmaceutical industry, which used to be quite similar to the chemical industry, or why it has been suggested that Japan may become the leading competitor of the United States in pharmaceuticals.[18]

In our taxonomy, however, only the industrial chemicals sector falls into the science-based category. Electrical machinery, communications and electronics, and drugs and medicines fall into the high-tech pattern. This may indicate that the way in which basic sciences are linked to technology development is different in these two patterns. To determine this, we need an empirical study that uses a more solid database than ours and addresses this issue more directly. Based on a study of citation data in U.S. patents, Francis Narin developed one measure of "science linkage," which he defined as the average number of patent references to scientific papers. These references, he felt, represented a company's need for scientific research.[19] Using the patents issued in the United States for the period 1985–1989, he measured the average number of references to scientific papers for each of fifty-five product groups. We selected only those product groups in which the number of patents examined was larger than ten (see Table 3-7). As shown in the table, the organic chemicals sector leads the survey in science linkage. The 177 patents in this product group cite on average 1.18 references to scientific papers. The drugs and medicines sector comes second, with 1.05 references per patent. The measured science linkage for the products of those industries, which we have identified as falling into the high-tech pattern, ranges from 1.05 to 0.14 references per patent. The science linkage for the products of those industries we have identified as part of the dominant-design pattern, however, on average falls below 0.12 references per patent.

In conclusion, even when patent citations are used, the high-tech pattern can be characterized as something between the science-based and the dominant-design patterns. Indeed, Narin's findings follow our classification of science-based, dominant-design, and high-tech industries almost perfectly, even though his analysis used a different database and models.

Competence-Destroying Discontinuities. After reviewing the results of Utterback and Suarez's analysis of the widespread applicability of the dominant-design theory, we have suggested that the theory is neither applicable nor effective in the high-tech industry. In fact, our empirical

TABLE 3-7
Measurement of science linkage in U.S. patents, 1985–1989

Product Group	Number of Patents	Science Linkage
Science-based pattern		
Organic chemicals	177	1.18
Inorganic chemicals	74	0.62
High-tech pattern		
Drugs and medicines	155	1.05
Machinery, excluding electrical	73	0.14
Communications and electronics	10	0.21
Scientific instruments	43	0.70
Dominant-design pattern		
Textile mill	19	0.10
Rubber, misc. plastic	101	0.12
Stone, clay, glass, concrete	11	0.09
Fabricated metal	18	0.03
Transportation equipment excluding air	13	0.07

Source: Data from F. Narin and D. Olivastro, "Status Report: Linkage between Technology and Science," *Research Policy* 21 (1992): 237–49. Reprinted by permission.

analysis forced us to establish a different taxonomy in which the electronics industry follows the high-tech pattern, while the automobile industry follows the dominant-design pattern. But what theory will explain this division?

In 1986, Philip Anderson and Michael Tushman demonstrated that technology evolves through periods of incremental change punctuated by technological discontinuities.[20] In doing so, they made a distinction between competence-enhancing and competence-destroying discontinuities. A competence-enhancing discontinuity builds on existing know-how in the industry, while a competence-destroying one renders existing knowledge obsolete. For example, while the advance in turbofan jet engines was built on prior jet competence, the skills of mechanical watch manufacturers or vacuum-tube producers were rendered irrelevant by Quartz watches and integrated circuits, respectively. Although Anderson and Tushman tried to relate this demarcation to the dominant-design theory, it may be more effective in our taxonomy. Used here, the high-tech industry follows the competence-destroying pattern, while industries in the dominant-design category follow the competence-enhancing pattern.

In fact, Anderson and Tushman's longitudinal study of cement (1888–1980), glass (1893–1980), and minicomputer (1958–1982) industries alludes to this possibility: in minicomputers, two competence-destroying discontinuities are identified for the five-year period from 1960 to 1965; for the cement industry, however, the majority of discontinuities are found to be competence-enhancing in the period from 1888 to 1972. Furthermore, unlike the cement or glass industry, the minicomputer industry does not confirm the hypothesis that competence-destroying discontinuities will be initiated by new entrants, while competence-enhancing discontinuities will be initiated by existing firms. On the contrary, early minicomputers were made by existing accounting machine and electronics manufacturers.

In our taxonomy, the manifest differences in an R&D program dynamics can explain why a dominant design is not applicable in industries following the competence-destroying pattern. Conversely, the existence of a dominant design can explain the dynamics inherent in an industry following the competence-enhancing pattern, i.e., the converging pattern of its freezing-rate function. Furthermore, the distinction in our taxonomy between the dominant-design pattern and high-tech pattern can explain the effectiveness of the distinction between competence-enhancing and competence-destroying discontinuities. In other words, our empirical taxonomy helps to make a discussion of competence-destroying discontinuities less tautological.

R&D Intensiveness. Various administrative branches of government in various countries have adopted an empirical approach to defining high-tech industries. One well-known definition, for example, is the one used by the Bureau of Economic Analysis in U.S. Department of Commerce.[21]

Based on the three or four digits of the Standard Industrial Classification Code (SIC), the bureau defines high-tech industry as any industrial sector that satisfies one of the following two conditions: (1) the percentage of the sector's R&D expense in its value-added output is larger than 10 percent; or (2) the percentage of the sector's number of scientists and engineers in its total employment is larger than 10 percent. Using these criteria, the bureau selected the following products as high-tech products: drugs (SIC 283), office, computing, and accounting machines (357), electrical machinery (36), aircraft and parts (372), and guided missiles and space vehicles (376). In its competitive assessment study, however, the bureau selected the following industries as high-tech industries: drugs and medicines, business machines and equipment, computers, electrical and electronic machines and equipment, telecommunication equipment, electronic components, consumer electronics, jet engines, aircraft, and scientific instruments. In Japan, MITI, using the Standard International Trade Classification (SITC), selected the following products as high-tech products: machine-tools (SITC

TABLE 3-8
Coincidence with selections by government agencies

High-Tech Pattern	Department of Commerce (U.S.)	MITI (Japan)
Drugs and medicines	drugs and medicines	
Ordinary machinery		machine tools
Electrical machinery	electrical and electronic machines and equipment	
	business machines and equipment	
Communications and electronics	telecommunication equipment; computers; consumer electronics	automatic data processing machines and units thereof; videotape recorders
Precision equipment	scientific instruments	
—	aircraft; jet engines	aircraft and associated equipment

Source: U.S. Department of Commerce, *An Assessment of U.S. Competitiveness in High Technology Industries*, Washington, D.C., 1983; and MITI, *White Paper of International Trade* (in Japanese) (Tokyo: Nihon-Toukei-Kyoukai, 1984).

736), automatic data processing machines and units thereof (7521, 7522, 7523, 7528), transistors and similar semiconductor devices (7764); aircraft and associated equipment (792), and videotape recorders.[22]

Although the selection procedures adopted by these two government agencies are different from those used in our study, the industrial sectors we have classified as high-tech coincide surprisingly well with those selected by these agencies with the exception of the aircraft industry, which we have included in the other transportation equipment sector (see Table 3-8).

It should be noted, however, that there is an obvious discrepancy between the various agencies' selection procedures and the results of these procedures. For example, industrial chemicals is not identified as a high-tech industry by either the U.S. government or the Japanese government even though it satisfies the two previously described conditions. Both governments, however, have not explained why it is excluded.

For both government agencies, there is no essential difference between the definition of R&D-intensive industry and high-tech industry.[23] The definitions used by these agencies seem to say that high-tech industry is nothing more than a highly R&D-intensive industry. However, R&D-intensive industry and high-tech industry are not the same because they are derived from different policy contexts, i.e., high-tech industry is discussed in terms of its competitiveness, while R&D-intensiveness is framed in terms of the government's support of basic science.

In our taxonomy, on the other hand, R&D-intensiveness is not exclusive to one pattern. High-tech industries and science-based industries can be R&D-intensive. In other words, these two concepts are orthogonal with each other. Unless a distinction is made between these concepts, policies that are extensions of policies applied to R&D-intensive industries may be applied inappropriately to high-tech industry. Policy planners may be told, for example, to spend more money on R&D and recruit more scientists and engineers to promote high-tech industry, or it may be suggested that government should support basic research because private companies cannot afford to any more.

MANAGING HIGH-TECH COMPETITION

The differences in R&D behavior between high-tech, dominant-design, and science-based industries have substantial management implications. Aoki's economic model, which was discussed in chapter 1, distinguishes between a hierarchical coordination mode and a horizontal coordination mode. The hierarchical mode is characterized by the separation of planning and implementation based on economies of specialization. The horizontal mode is characterized by horizontal coordination among operating units based on knowledge-sharing.[24] According to Aoki, the hierarchical mode may be superior in achieving the organization's goal in either stable or extremely volatile planning environments. In intermediate situations, however, where external environments are continually changing but not drastically, the horizontal mode is superior.

Now that we understand the structural differences in R&D behavior that exist among the three types of industries, we can extend Aoki's theories to the strategic decision level. We will try to do this in terms of the coordination between R&D management and corporate management.

Managing the Science-Based Pattern

In a dominant-design industry, a key technology (such as continuous casting in the steel industry) is more or less given. Therefore, the problem is how and when to implement capital investment commercializing the key technology to maximize its rate of return. In this situation, we can separate R&D management from corporate management. In fact, the development of appropriate procedures for coordination among operational decision units is vital in these industries. The *kanban* (just-in-time) system at Toyota (described in chapter 1) facilitates a smooth flow among different plants; an integrated-engineering control-room locates and solves cross-shop problems in a manufacturing environment based on the continuous casting.

In a science-based industry, however, R&D management is closely linked to a company's survival. In essence, this requires management of the unmanageable because the process of basic science is essentially random. One management concern, therefore, is how to hedge against high risks. The only way this can be done is to build a strong financial base to support as many basic-research programs as possible. By doing so, R&D management can be separated from corporate management. American and European chemical giants seem to be following this type of strategy. Table 3-9 compares the financial strength of a variety of chemical companies in terms of total assets, stockholders' equity, and their equity ratio to assets as a measure of financial independence.

As the table indicates, German chemical giants are largest on average in the amount of total assets, and their stockholders' equity ratio to total assets is on average 40 percent, the highest among three countries. American chemical giants are the second both in total assets and stockholders' equity ratio to assets. The average amount of total assets among nine of the largest Japanese chemical firms is equivalent to one-fourth of the German figure and one-third of the U.S. figure, while the subtotal of total assets is not much lower than the subtotals for German and U.S. firms. The ratio of stockholders' equity to total assets, however, is 25 percent on average among the nine Japanese firms. This figure is far below 40 percent, the German figure, and below 30 percent, the U.S. figure. As we noted in chapter 1, government intervention so fragmented Japanese chemical firms that they could not exploit economies of scope in R&D. Therefore, a major reorganization of Japanese chemical industry will occur sooner or later. In fact, recently, two Japanese chemical giants announced plans to merge.

In high-tech industry, the situation is quite different. In this industry, organized and targeted basic research is more important than random support of basic research. Therefore, R&D management cannot be separated from corporate management, and the two types of management should be consolidated. For this to occur, some organizational innovation is needed.

TABLE 3-9
International comparison of financial strength among major chemical companies

Country/ Companies	Total Assets (A) ($million in 1992)	Stockholders' Equity (B) ($million in 1992)	Equity to Assets (B/A)
United States	(17,293)		(30%)
DuPont	38,870	11,528	30%
Dow Chemical	25,360	8,064	32
Monsanto	9,085	3,005	33
Union Carbide	4,941	1,238	25
SUBTOTAL	78,256		
Germany	(23,585)		(40%)
Hoechst	22,869	8,585	38
BASF	24,147	9,035	37
Bayer	23,747	10,838	46
SUBTOTAL	70,763		
Japan	(5,824)		(25%)
Asahi-Kasei	8,842	3,106	35
Mitsubishi-Kasei	7,631	2,125	28
Sumitomo Chemical	6,825	1,241	18
Showa Denko	6,583	1,520	23
Ube Industries	5,524	743	13
Mitsui-Toatsu Chemicals	5,086	1,159	23
Mitsubishi Petrochemical	4,723	1,692	36
Toyo Soda	3,864	687	18
Mitsui Petrochemical	3,342	1,184	35
SUBTOTAL	52,420		

Source: For data on Japanese companies, *Japanese Companies Handbook* (in Japanese) (Tokyo, Toyo-Keizai-Shinpohshya, 1993); for data on U.S. and German companies, *Nikkei Annual Foreign Corporation Reports* (in Japanese) (Tokyo: Nippon-Keizai-Shimbunshya, 1994).

Note: Country's averages are in parentheses.

Managing Diverging Patterns

The discernible diverging pattern of the high-tech competition revealed in the detailed study of DRAM development earlier in this chapter pointed out that corporate decisions on investment cannot be made on the basis of the rate of return. Instead, they must be made according to the principle of *surf-riding*. In other words, companies either invest in successive waves of innovation, or they are left behind by competitors. Investment must continue just to stay in the market.[25]

One organizational response to this pattern is the concurrent pursuit of sequential generations of a product line and its associated technol-

FIGURE 3-7
Sequential versus overlapping phases of development

Type A

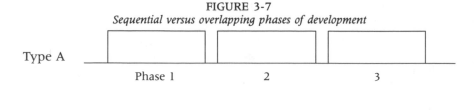

Phase 1   2   3

Type B

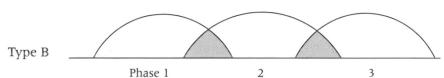

Phase 1   2   3

Source: Adapted from H. Takeuchi and I. Nonaka, "The New Product Development Game," *Harvard Business Review* 69, no. 1 (1986): 139. Adapted and reprinted by permission.

ogy. Phase management in Japanese companies tends to be overlapping rather than sequential (see Figure 3-7). The sequential approach (Type A) is used by a number of U.S. companies. Under this system, new product development moves through different phases step-by-step. In the overlapping approach (Type B), overlapping occurs at the interface of adjunct phases.[26] In recent years, overlapping has become common in Japanese companies in every industrial sector, especially the automobile industry.[27]

Although overlapping is necessary in high-tech industry, it is not enough. While the use of overlapped development may result in a shorter development time and an increase in the interaction between the groups responsible for different phases of the innovation cycle, we need something more to manage the high-tech pattern.

D. Okimoto and Y. Nishi have described how one Japanese semiconductor firm is managing the diverging pattern of DRAM development.[28] They note that the system is designed to organize engineers from development and manufacturing into a series of small research teams. These teams are used to sustain research simultaneously for consecutive DRAM generations. For example, a team was started on the 64K DRAM, and two years later, while the 64K project was still in progress, another team began to work on the next generation, the 256K DRAM. Four years later, when the 64K DRAM was being mass produced and prototypes for the 256K were being developed, work began on the 1M DRAM. Following the completion of the 64K project, the 64K team shifted over to work on the 1M DRAM. Their move was followed by the 256K team's shift to the 4M DRAM project (see Figure 3-8).

Many blue-chip Japanese electronics giants possess the technical manpower and financial resources necessary to support a multi-layered structure of research labs. These firms may have a central research laboratory (CRL), where basic and advanced development research is conducted; divisional labs (DLs), where product development and process technology take place; and factory engineering labs (FELs), where manufacturing specifications and processes are ironed out. The CRL is funded largely by corporate headquarters and is in charge of research projects lasting more than five years. DLs are responsible for projects taking three to five years, and their funding is underwritten by each division. FELs are responsible for R&D with a time frame of less than three years, and their funds come out of the budgets for each factory.

Located at the nodal point of contact between the CRL and the FEL, the DLs are the primary mechanism for technology transfer from theoretical inquiry to commercial application. The DLs are expected to turn out product prototypes that can be debugged and made in small volumes before they are turned over to the factories for mass manufacturing. For the commercialization cycle to move forward quickly, the DLs must be able to complete the nitty-gritty tasks of designing new products and ironing out all the problems of process technology within a tight schedule.

Toshiba was the first firm to devise this widely emulated organizational approach to R&D. Using the three-layered structure, Toshiba developed the system of "three generations parallel sequencing" of DRAM development (see Figure 3-9). For Toshiba, the CRL (USLI research lab) develops a fundamental technology, the DL (semiconductor device engineering lab) brings this technology into the prestage of mass manufacturing, and the FEL (engineering department in memory division) establishes a mass-manufacturing system. Toshiba's engineering center for semiconductor systems is responsible for marketing, application, and testing.

In the current state of the art for DRAM, 16M DRAM is now mass produced, 64M DRAM is being shipped for sample testing in certain uses, and 256M DRAM is at the development stage. Reflecting these chip-manufacturing dynamics, the CRL is responsible for 256M, the DL for 64M, and the FEL for 16M DRAM. In this approach, responsibility continuously circulates among these three institutions. The DL, for example, is divided into two research teams, one for 16M and the other for 64M. When the team for 16M finishes its development, it will become responsible for 256M DRAM.

FIGURE 3-8

Concurrent pursuit of sequential generations of DRAMs

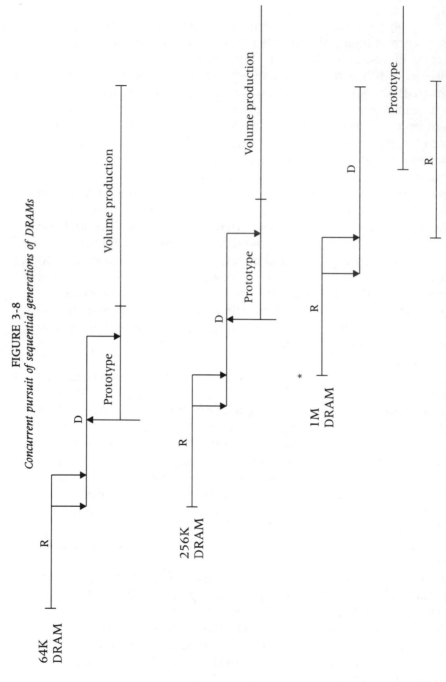

Source: D. Okimoto and Y. Nishi, "R&D Organization in Japanese and American Semiconductor Firms" (paper presented at the Stockholm School of Economics Conference, Stockholm, Sweden, 1991). Reprinted by permission.

134

FIGURE 3-9
Three-layered structure for semiconductor development

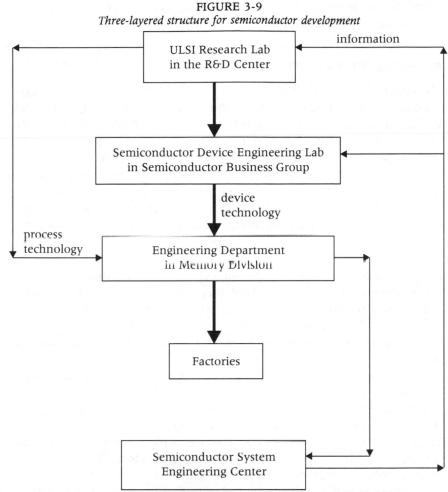

Source: *Nikkei-Sangyo* newspaper (in Japanese), 8 October 1991. Reprinted by permission.

Toshiba's parallel sequencing system has come about because progress in miniaturization, with all its associated process and tool changes, is more rapid than the time required to bring up a new process and obtain commercially useful yield. While integrated circuit densities double every eighteen months (linear dimensions are halved every three years), it takes five or six years from research to first commercial production for each of these steps. Under these circumstances, phase management should be organized as multiple, interlaced spirals with each function (research, development, production engineering, market entry) taking place concurrently, and progress in each function proceeding faster than the spiral can be completed.

Managing Interindustry Competition

Through technological diversification, a high-tech company can defend itself against interindustry competition. As described in chapters 1 and 2, SEI's diversification strategy was designed to ensure that its strong position in communication cable manufacturing was not threatened by unexpected innovations outside the industry. It was this strategy that led SEI to view the emergence of optical-fiber as a threat and an opportunity, and the company quickly diversified into this area.

Technological diversification is not only a good defensive strategy but also a way to spot innovations worthy of investment. Management's biggest dilemma, however, lies in deciding what technologies to focus on and where to look. Because it must set realistic boundaries, a high-tech company has to follow a technology concentration strategy while looking for worthwhile innovations. In this context, NEC Corporation tries to strike the optimal balance beween diversification and concentration strategies. Its diversification is in the markets, and its concentration is in its computation and communications technology (see Figure 3-10).[29]

This strategy has evolved from NEC's belief that R&D productivity depends upon how broadly technology is applied, while R&D activity is constrained by the number of scientists a company can employ. In other words, R&D productivity depends on economies of scope. Following this belief, NEC sought a strategic planning system to identify the minimum number of technologies needed to capture the maximum number of markets with growth potential.

In order to realize this strategy in setting its corporate R&D agenda, NEC developed a well-planned, formalized program called a "core technology program" in 1975. In this program, considerable time and effort are devoted to understanding and choosing the set and subsets of core technologies. For two years every decade, a group of fifty middle and senior managers from marketing, operations, and research and development collectively analyze the company's overall technological needs for the next ten years. NEC's CRLs work as catalysts and mediators to focus the discussion. In 1975, the company defined twenty-seven core technologies; by 1990 the number had increased to thirty-four. In coming up with a set of technologies, the fifty participants sift through a broad assortment of market and technological information that, ordinarily, none of them would bother to look at. The job not

FIGURE 3-10

NEC's concentration and diversification strategy based on its C&C logo

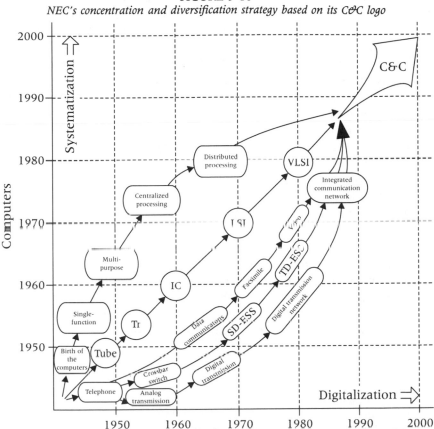

Source: Michiyuki Uenohara, NEC (paper presented at the Seminar on the Development of High Technologies in the U.S.A. and Japan: Comparative Evaluation and Policy Implications, John F. Kennedy School of Government, Harvard University, 1992). Reprinted by permission.

only requires formal information-gathering capability but also instills an appreciation for looking outside the company for new ideas. Thus, this method may implicitly induce monitoring hidden enemies.

We can view NEC's system in the broader context of R&D management. First, the purpose of this strategic planning system is to focus enough resources on a defined set of technologies to reduce the risk of technical failure in any one technology to an acceptable level. The residual risk is market risk, but this will be minimized if the planning process has successfully identified the most attractive markets. There are, of course, constraints on market choice because the strategy requires that each of the markets served reinforce the technical strategy.

Second, this system helps to overcome the inherent difficulties in technology transfer from the CRL to product divisions and helps to develop CRL relationships with product divisions, thus giving these divisions a stake in CRL's relevance and success. By acting as a catalyst and mediator in a core technology program, the CRL tries to meet the needs of the company's business divisions by preempting interindustry competition in forecasting the key technologies each business has to develop and in defining the technical barriers to entering new markets. The CRL has to develop new key technologies before the business divisions define their needs.

Third, this system is a way to consolidate research management with business management, a basic problem in the high-tech industry. Corporate management presses for more efficient R&D, but research management must focus on effective R&D. R&D effectiveness is, however, highly dependent on the way it is managed, and technological value is created mostly by management. We should distinguish an efficient R&D management style that maximizes technical output from an effective one that places a priority on selection of technical goals and building relationships that ensure a successful project.

CONCLUSION

Through two case studies, we have formulated two kinds of high-tech competition: interfirm competition following the diverging pattern and interindustry competition. In order to understand where these phenomena come from, we constructed a state transition model of an R&D program.

This model formulation was based on a freezing-rate function. By differentiating three types of freezing-rate functions, we could make a distinction among dominant-design, science-based, and high-tech patterns. A statistical fitting to the R&D database collected by the Japanese government identified the unique nature of high-tech industry: an R&D program may be frozen even after it has entered the development state. On the basis of our quantitative analysis, we established an empirical taxonomy of R&D activities, which included several independent findings concerning the dynamic nature of technical changes.

Since the risk of being frozen comes from competition with both visible and invisible rivals, we focused on how to manage these unique features of competition: how to deal with a diverging pattern of competition among companies in the same industrial sector; and how to concentrate R&D resources so that technical surprises can be monitored and avoided. All of the Japanese practices that we have described in this chapter, however, are fairly new. Thus, we have to ask if they are sustainable. It may be necessary to invent a new management method to cope with high-tech competition in a sustainable manner.

NOTES

1. J. Utterback and W. Abernathy, "Dynamic Model of Product and Process Innovation," *Omega* 3, no. 6 (1975): 639–56.

2. R. Nelson, "The Co-Evolution of Technology, Industrial Structure, and Institutions," *Industrial and Corporate Change*, forthcoming.

3. J. E. Tilton, *International Diffusion of Technology* (Washington, D.C.: The Brookings Institution, 1971).

4. Masahiko Aoki, "Toward an Economic Model of the Japanese Firm," *Journal of Economic Literature* 28 (1990): 8.

5. Michael Gort and Steven Klepper, "Time Paths in the Diffusion of Product Innovations," *The Economic Journal* 92 (1982): 630–53. Michael Tushman and Philip Anderson, "Technological Discontinuities and Organizational Environments," *Administrative Science Quarterly* 31 (1986): 439–65.

6. James Utterback and Fernando Suarez, "Innovation, Competition, and Market Structure," *Research Policy* 22 (1993): 1–21.

7. Ibid., 15.

8. V. Nicholas, "Thinking Ahead: A Survey of Japanese Technology," *The Economist* 313, no. 7631 (1989): 2–8.

9. D. Okimoto and Y. Nishi, "R&D Organization in Japanese and American Semiconductor Firms" (paper presented at the Stockholm School of Economics Conference, Stockholm, Sweden, 1991).

10. National Research Council, *International Competition in Advanced Technology: Decision for America* (Washington, D.C.: National Academy Press, 1983), 28–38.

11. F. Kodama et al., "The Innovation Spiral: A New Look at Recent Technological Advances" (paper presented at the Second US-Japan Conference on High Technology and the International Environment, Kyoto, Japan, 1986).

12. R. Evenson and Y. Kislev, *Agricultural Research and Productivity* (New Haven: Yale University Press, 1975), 377–85.

13. Prime Minister's Office of Japan, Statistical Bureau, *Report on the Survey of Research and Development* (in Japanese) (Tokyo: Nihon-Toukei-Kyoukai, 1970–82).

14. F. Kodama and Y. Honda, "Research and Development Dynamics of High-Tech Industry," *The Journal for Science Policy and Research Management* 1, no. 1 (1986): 65–74.

15. N. Rosenberg, *Perspectives on Technology* (Cambridge: Cambridge University Press, 1983).

16. U.S. Department of Defense, NASA, *Phased Project Planning Guideline*, NHB 7121.2 (Washington, D.C., 1968).

17. K. Pavitt, M. Robson, and J. Townsend, "Technological Accumulation, Diversification and Organization in UK Companies," *Management Science* 35 (1989): 81–99.

18. M. Dibner, "Biotechnology in Pharmaceuticals: The Japanese Challenge," *Science* 229 (1985): 1230.

19. F. Narin and D. Olivastro, "Status Report: Linkage between Technology and Science," *Research Policy* 21 (1992): 237–49.

20. M. Tushman and P. Anderson, "Technological Discontinuities and Organizational Environments," *Administrative Science Quarterly* 31, (1986): 438–65; see also P. Anderson and M. Tushman, "Technological Discontinuities and Dominant Designs: A Cyclical Model of Technological Change," *Administrative Science Quarterly* 35 (1990): 604–33.

21. Office of Technology Assessment, *Technology, Innovation, and Regional Economic Development* (Washington, D.C., 1982); U.S. Department of Commerce, *An Assessment of U.S. Competitiveness in High Technology Industries* (Washington, D.C., 1983).

22. Ministry of International Trade and Industry, *White Paper of International Trade* (in Japanese) (Tokyo: Government Printing Office, 1984), 130.

23. National Science Foundation, *Science Indicators* (Washington, D.C.: U.S. Government Printing Office, 1982).

24. M. Aoki, "Toward an Economic Model of the Japanese Firm," 8.

25. F. Kodama and Y. Honda, "Research and Development Dynamics of High-Tech Industry." See also F. Kodama, "How Research Investment Decisions Are Made

in Japanese Industry," in *The Evaluation of Scientific Research*, ed. D. Evered and S. Harnett (New York: John Wiley & Sons, 1989), 201–14.

26. H. Takeuchi and I. Nonaka, "The New Product Development Game," *Harvard Business Review* 64, no. 1 (1986): 137–46.

27. Ken-ichi Imai, "The Japanese Pattern of Innovation and Its Evolution," in *Technology and the Wealth of Nations*, ed. Nathan Rosenberg, Ralph Landau, and David Mowery (Stanford: Stanford University Press, 1992): 233; and K. Clark and T. Fujimoto, *Product Development Performance: Strategy, Organization, and Management in the World Auto Industry* (Boston: Harvard Business School Press, 1991), 205–45.

28. D. Okimoto and Y. Nishi, "R&D Organization in Japanese and American Semiconductor Firms."

29. L. Branscomb and F. Kodama, *Japanese Innovation Strategies: Technical Support for Business Visions* (Lanham, Md.: University Press of America, 1993).

4

Product Development
From Pipeline
to Demand Articulation

Recent events have shown that scientific leadership does not necessarily translate into industrial or product leadership. Therefore, we need to consider the connection between science and product.[1] Usually this connection is described as a type of *pipeline progression* in which a new technology emerges successively from basic research, applied research, exploratory development, engineering, and manufacturing.[2] R. Gomory has called this progression the *ladder process:* the step-by-step reduction of new scientific knowledge into a radically new product.[3] In the ladder process, according to Gomory, a new technology dominates, and a product is created around it. The customers' needs are taken for granted.

A classic study, the *SAPPHO* (Scientific Activity Predictor from Patterns with Heuristic Origin) Project, conducted by SPRU (Science Policy Research Unit) in the United Kingdom, has pointed out, however, that understanding users' needs is the most critical factor for successful innovation.[4] This does not mean that a firm should simply rely on market research to achieve successful innovation. Gomory notes that the Japanese are highly responsive to markets not through careful market research, but in a pragmatic way.[5] The Japanese company, according to Gomory, gets the product out fast, finds out what is wrong with it, and rapidly adjusts. In other words, a shorter development and manufacturing cycle can beat market research every time.

We argue, however, that the emergence of high technology makes both the pipeline view and the incremental improvement approach inadequate in themselves. High-tech products frequently undergo long periods of incremental improvement with gradually declining benefit/cost ratios that are punctuated by bursts of radical technical changes, usually traceable to basic research, that remove fundamental barriers to further incremental improvement. Furthermore, in developing high-tech products, it is important to identify "virtual markets." These are hypothetical markets in the sense that the market for a product is not articulated until the technology is created. "Virtual markets" arise from perceptions of social and economic need and, in that sense, are "socio-technical" rather than simply possibilities suggested by technology alone.

Indeed, the emergence of high technology has caused a drastic change in the government policy arena. For governments, the key issue is no longer how to break through technological bottlenecks, but how to put existing technology to the best possible use. Accordingly, technology policy, which traditionally has emphasized the supply side of technology development, now must work from the demand side. But the process of technology development is not as simple as the traditional demand-pull versus technology-push. It lies in between and requires a lot of feedback. Given this, we have to find a new and accurate way of describing the dynamic process of technology development. We have to give science policy administrators and research managers a vocabulary and a framework for talking about the choices they must make in the high-tech environment.[6]

To do so, we will introduce the concept of "demand articulation," a sophisticated translation skill that converts a vague set of wants into well-defined products. Articulating demand is a two-step process: market data must be integrated into a product concept, and the concept must be broken into development projects. Potential demands are often derived from virtual markets. The fact that the technology is still considered exotic should not be a deterrent in setting development agendas.

To confirm that articulating a potential demand is critical to successful product development at the company level, we will consider two case studies of product development in Japanese high-tech companies. Then, in order to demonstrate the effectiveness of demand articulation in government policies and its universality, we will review the history of the development of IC technology, first in the U.S. defense sector, and then in Japanese government-sponsored research consortia. In doing so, we will arrive at the concept of a national system of demand articulation, which is analogous to the often discussed concept of a national system of innovation.[7]

To quantify the demarcation between the pipeline view and demand articulation, we will make a mathematical model of participation behavior in two different kinds of collective research, an international collaborative research project organized by the International Energy Agency (IEA) and NTT-initiated collective research programs among communication equipment manufacturers. The IEA project exemplifies the pipeline view of technology development. The NTT programs exemplify the demand articulation view of product development.

Finally, we will discuss the management implications of demand articulation in two parts. The first part will concern the public management of research consortia. Here, we will review the process of social learning to understand the economic and technological rationale behind collective research in which rival firms participate.[8] The second part concerns international business implications. In this part, we will discover how a supplier can capitalize on the demand articulation capabilities of its customer company, even if the company is located in other countries.

DEMAND ARTICULATION

In order to succeed at product development, a company must have good marketing, a good understanding of economics, and good technologies. The economics of innovation stress the cost-effectiveness side of innovation, i.e., innovation occurs when it pays. Since this framework of analysis is static and retrospective, it does not help in the development of a technology strategy, which must address the dynamic aspects of innovation.

Some economists have noted the intrinsic dynamics of technology development. Nathan Rosenberg, for example, has concluded that *backward linkage* has been an enormously important source of technical change in the Western world.[9] He argues that the ordinary messages of the marketplace are not specific enough to indicate the direction in which technical change should be sought. Therefore, he concludes, there must be forces outside the marketplace that point in certain directions. Rosenberg suggests that bottlenecks in connected processes and obvious weak spots in products present clear targets for improvement. These become the *technological imperatives* that guide the evolution of certain technologies. Rosenberg's approach is, of course, historical, and his concept of technological imperatives is based on long-term, macrolevel phenomena of technical change. To analyze the product development process, however, we need a microlevel version of his concept.

While many articles in the market research field have covered the market for technology and pointed out the importance of understanding market needs, few have mentioned technology development and the market. Outside of market research articles, there has been some work done on technology development and the market. Eric von Hippel of MIT, for example, argues that users who understand the needs of the market usually develop the technology first.[10] He has proposed a *customer-active paradigm* of technical innovation, but his analysis is limited to innovation in industrial goods, such as scientific instruments and semiconductor process equipment. In the area of consumer goods, it is hard to imagine end users as a source of innovation. Here too, we need to extend his concept to the development process of consumer goods.

From the technologists' viewpoint, S. Kline argues that innovation can be interpreted as a search and selection process among technical options.[11] The sample population from which technical options can be drawn, however, varies over a wide spectrum of sources of innovation. An option may be drawn from existing technical collections in which marketing and economic research are predominant. It may also be drawn from a pool of scientific knowledge, an approach advocated by scientists and technologists in search of a *linear* progression model of innovation. Between these two extremes, however, there is a wide range of technology development that might be best described as "targeted technology development."

In this kind of technology development, Nelson and Winter's "alternatives out there waiting to be found" is somewhat forced.[12] The most important element in targeted technology development is the process in which the need for a specific technology emerges and R&D effort is targeted toward developing and perfecting it. This is what we call *demand articulation*. The word "articulate" has two conflicting meanings: (1) to divide into parts and (2) to put together by joints.[13] Thus, the word encompasses two opposite concepts: analysis (decomposition) and synthesis (integration). In fact, both are necessary in technology development, and the heart of the problem concerning technology development is how to manage these conflicting tasks. Therefore, I define demand articulation as a dynamic interaction of technological activities that involves *integrating* potential demands into a product concept and *decomposing* this product concept into development agendas for its individual component technologies.

Sometimes potential demands are derived from distant human wants—a home-use video-tape-recorder (VTR) that the average family can afford. At other times, demands are derived from virtual markets—the Japanese language word processor, which arose from fears that the Japanese language itself would become a bottleneck in office automation.

In setting development agendas, one should not exclude emerging technologies. As described in chapter 2, Sharp was quick to identify liquid crystal display as a promising technology when the company was translating customer's desire for a more powerful, thinner electronic calculator into specific R&D projects. The fact that the technology was still considered exotic was not a deterrent. Instead, Sharp saw LCDs as a way to solve specific technical problems and change the rules of competition in the marketplace.

The two case studies that follow will show that demand articulation is critical for successful product development in Japanese high-tech companies. The process of integrating potential demands into a product concept is illustrated by the development of Japanese-language word processors. The process of decomposing a product concept into development agendas for component technologies is shown through the development of the home-use VTR.[14] Both of these cases demonstrate that a long-term commitment to a specific product concept makes consistent demand articulation possible. This, in turn, permits the successive development of new technologies to meet specific technological needs. In retrospect, it is clear that these achievements could not have occurred as mere extrapolations of existing technologies.

Japanese Language Word Processors

In the development of Japanese-language word processors, careful investigations into a cultural factor (the complexity of the Japanese written language), made demand articulation possible. Because the technology was directly related to the culture of the country, needs had to be carefully articulated and adequate time had to be devoted to fundamental research.

The Japanese language uses a mixture of the 50-character *kana* alphabet and over 3,000 Chinese *kanji* characters. The keyboard of a traditional Japanese typewriter—actually a plate printed with rows of tiny characters and a mechanical arm with a pointer to select them—contains more than 3,000 characters that the operator must find and punch in one by one. Not surprisingly, only trained specialists are able to operate Japanese typewriters. Thus, the diffusion of Japanese typewriters was far behind that of the American and European machines, and it was feared that this would limit office automation in Japan.

The first attempt at a Japanese-language word processor, which also used a plate printed with individual characters, ended in failure. Manufacturers were forced to return to fundamental research, starting with the basic principles of linguistics. After ten years of wide-ranging research at Toshiba's laboratories, a new method was devised in which *kana* letters could be converted to *kanji* characters. Working from a keyboard of 48 *kana* letters, an operator can key in text as it is pronounced, while a computer, programmed with a dictionary of Japanese vocabulary and grammar, automatically converts the letters into *kanji* characters where necessary. This conversion is illustrated in Figure 4-1.

During development, research was also conducted in related technologies, such as dot-matrix printers and cathode-ray tubes for display. In choosing a printer technology, it was clear that the minimum requirement for a Japanese-language word processor was *24/24 dots*, not *16/16 dots*, which was available at the time. In order to create the printer, the diameter and arrangement of the pin on print head and on materials was researched. As a result, the new wire dot printer was developed. One reason for developing this printer was the difficulty of making changes in official documents. At the time, the resolution capacity of cathode-ray tubes for display was *600-700 lines*, but A4 size paper and forty characters of 24 dots per line requires 1,000 lines. This problem was solved by enhancing the horizontal resolution of the cathode-ray tube.

In the seven years since the development of Japanese-language word processors, their price has been reduced by 98 percent and their size has been significantly reduced. Thus, they are now widely used in offices and in homes. Indeed, the market for them has grown to 2 million units annually.

The development history of culture-specific products is interesting because development must be conducted domestically and cannot draw on a foreign technological reserve. Moreover, once goals have been set, the technology has to be improved until these goals are fully met.

Home-Use VTRs

Demand articulation became possible in the case of home-use video tape recorders (VTRs) because of a long-term commitment on the part of manufacturers to a specific product concept. Ever since the appear-

FIGURE 4-1
Illustration of conversion from kana to kanji

○はし (Hashi) ⟶ 端 (End), 箸 (Chopstick), 橋 (Bridge)

○こうしょう (Koushou) ⟷ 交渉 (Negotiation), 高尚 (Nobleness), 鉱床 (Deposit), 公称 (Proper name), 考証 (Investigation)

こうしょう-な (koushou-na) ＝ 高尚な (Noble)

○はし-を-わたる (Hashi-wo-wataru) ＝ 橋を渡る (Cross a bridge)

はし-を-つかう (hashi-wo-tukau) ＝ 箸を使う (Use a pair of chopsticks)

Source: Sei-ichi Takayanagi, Toshiba (paper presented at the Seminar on the Development of High Technologies in the U.S.A. and Japan: Comparative Evaluation and Policy Implications, John F. Kennedy School of Government, Harvard University, 1992). Reprinted by permission.

ance of VTRs for the broadcasting industry, Japanese manufacturers had been aware of a future for home-use VTRs. In choosing the technologies to use in developing industrial VTRs, these manufacturers took into account the eventual development of home-use VTRs. Moreover, their successes were not the result of choices among existing technologies, but the result of basic research into the fundamental format for VTRs and component technology.

In the video tapes used by the broadcasting industry, several perpendicularly rotating magnetic heads record the video signals for each frame in track lines that run perpendicular to the length of the tape. The movement of the magnetic heads causes a vibration of the tape as it runs through the machine, resulting in image jitter on the TV screen. Controlling the vibration was a key technological challenge.

For home-use VTRs, however, it was more imperative to find a way to use narrower video tape without splitting the picture, which would cause the screen image to fluctuate. Thus, in 1955, Norikazu Sawazaki of Toshiba invented the helical scanning system: an innovative means of recording signals diagonally across a narrow tape with a helically rotating magnetic head so that a field video signal could be put on a single recording track line without jitter.[15] The helical scanning system is compared to the conventional one in Figure 4-2. During the mid-1970s, Toshiba was attempting to develop another system of horizontal recording, although the home-use VTR market was beginning to grow. Sony, on the other hand, had made a long-term commitment to home-use VTRs and used the helical scanning system. Sony succeeded in articulating the market demand with the helical scan technology, and the standardization process began before Toshiba's new system was perfected. Today, the helical scanning system is widely used in almost every VTR.[16]

For color VTRs, in which the density of colors is expressed through the amplitude of the subcarrier of color signals and color tones are expressed through the phase of the carrier, a slip in time can cause a slip in the phases. In the industrial-use VTRs developed by Ampex, this problem was corrected through massive electronic circuitry. Massive electronic circuitry, however, was not feasible for home VTRs. Sony reduced the frequency of the carrier to one-sixth, a level at which small time slippages would not affect color.

The 8mm VTR came about as a result of the demand for a miniaturized and lighter machine. The fundamental technology for it was the development of a higher density recording medium. Iron oxide tapes, which had been used for the last thirty years, were reaching physical limits. On the other hand, metal tape technology was not yet realized.

FIGURE 4-2

Comparison of helical scanning with conventional system

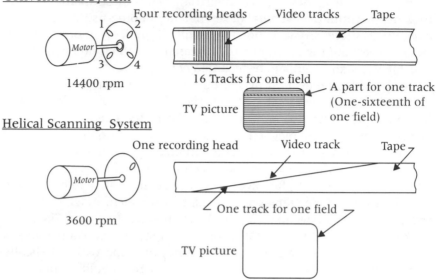

Source: N. Sawazaki et al., "A New Video-tape Recording System," *Journal of the Society of Motion Picture and Television Engineers (in Japanese)* 69 (1960): 868–71. Reprinted by permission.

In order to utilize metal tape in a VTR, various technologies had to be developed, such as a high-quality head for the metal tape, a smooth and rust-free surface for the metal tape, and a low temperature affinity of the tape base. In the work to develop a head, for example, a metal head was used first, then a ferrite one. The magnitude of the magnetic field of the ferrite head was far below the level needed for a metal tape. Therefore, a composite head was developed. Metal was used only for the critical part of the head gap, and ferrite was used for the total magnetic circuitry, because its reproduction characteristics were better.

Prerequisites for Demand Articulation

The development of Japanese-language word processors and home-use VTRs reveals that there are at least three prerequisites for successful demand articulation. *First,* management must take a long-term view of product development. This means a long-term commitment to providing stable and adequate financial and human resources for research and development.

Second, a given industry's capacity for demand articulation depends on the technological level of related industries. The more competent the industry as a whole, the higher the absorption rate of technologies from other industries. Thus, all industries involved in a product's development must have a high level of technological capability before a high degree of demand articulation can take place.

Third, demand articulation requires brisk competition between companies, almost to the point of excess. Competition motivates companies to focus on the customers' demand. At the same time, the competitive environment spurs high-tech companies to experiment with alternatives that they might not explore if competition is less intense. Indeed, competition in technology development is ultimately competition over how skillfully demands can be articulated.

A NATIONAL SYSTEM OF DEMAND ARTICULATION

Two things make the concept of demand articulation unique. First, it can be extended into the public policy arena. Second, it is universal; it extends beyond national boundaries. Perhaps the best illustration of the applicability of demand articulation is the development of the greatest invention in the postwar history, the integrated circuit, which was begun in the U.S. defense sector and carried further in Japanese government-sponsored research consortia.

Prior to the development of integrated circuits, programs sponsored by the U.S. Department of Defense were driven by technology rather than the need for a technology. In the case of the integrated circuit, however, the U.S. government articulated and shaped the problem to which the innovative candidate technology needed to be addressed.[17] Ironically the two basic patents and key technological contributions that underlie the IC innovation were made by private companies with no government support. They came about, however, because the companies knew of the efforts and interests articulated by the military. Thereafter, the U.S. government played a major role in bringing the innovation to fruition.

In the transition from the defense to the civilian market, more specifically, from the prototype market through the military and industrial market to the consumer market, leadership in technology development shifted from the United States to Japan. When it became obvious that integrated circuit technology could be applied to home computers and consumer electronics, MITI decided to orchestrate the establishment of a research consortium, the *ERA* for *VLSI* development. Although the association included all five of Japan's major IC manufacturers at the time, it did not directly help these chip makers in developing manufacturing technologies. Instead, by gathering all the major chip manufacturers together in one place, the association was able to articulate the demand for manufacturing equipment and materials. None of the five IC manufacturers were equipment or silicon suppliers. Thus, an internationally competitive infrastructure was established.[18]

U.S. Defense Sector

While it is known that IC technology was developed in the United States through defense research, a detailed investigation into the development process will reveal the importance of demand articulation. A vaguely defined demand for national security did not propel the development of IC technology; it came about because the U.S. security policy was successfully translated into a technological concept.[19]

In the years immediately following World War II, U.S. security policy emphasized the sheer destructive power of nuclear weapons. The strategy is best illustrated by Secretary of State Dulles's emphasis on massive retaliation. By the mid-to-late 1950s, however, when the Soviet Union had developed its own nuclear weapons, the inadequacy of such a simple notion of destructive power strategy was increasingly recognized. With the intensification of the cold war, the U.S. defense strategy became one of deterrence, rather than retaliation. A RAND-based study emphasizing assured second strike forces, the loss of overseas bomber bases, and its own preferences, led the Kennedy administration to create a new strategy that was dependent upon precision in the delivery of nuclear weapons. The demand, therefore, was for technologies that would carry nuclear weapons quickly and accurately to their targets. Thus the development of missiles became paramount. Defense R&D then focused on the development of small reliable electronic circuitry to control the missiles. In this way, the defense strategy of deterrence was translated into the technological problems of miniaturization and higher reliability of electronic circuitry.

It soon became clear that these requirements could not be fulfilled by conventional vacuum tube technology or by transistors. In 1958, the U.S. Air Force suggested the use of molecular electronics. In brief, components using this technology would have various electronic functions without specifically fabricating such individual parts as transistors, diodes, capacitors and resistors. The material used would *simulate* the electronic function of oscillators and amplifiers.[20]

Responding to the concept, various laboratories across the country began R&D experiments, articulated by the military, and IC technology was born. It is worth noting that Texas Instruments (TI) and Fairchild Semiconductor, which made key innovations in IC technology, sought no support from the U.S. government for their early IC development work. On its own, TI, a major electronics components manufacturer, initiated an in-house program to seek some basic new directions. By mid-1958, TI was able to demonstrate the first IC. Electronic components were indivisibly *embedded* within semiconductor materials through the use of photolithographic techniques. In late 1958, TI demonstrated its device to the air force, and in mid-1959, the air force awarded a two-and-a-half-year contract to TI at $0.5 million per year for the development of various IC devices.

As the IC R&D base proved itself, the translation of laboratory achievements into components and equipment became important. In early 1961, another air force contract for $2.8 million required TI to build a computer using IC components. Later that year, TI displayed its first computer. The system offered impressive advantages over its predecessors and served as a showcase for the potential utility of integrated circuits. Indeed, it was the first indication that the promise of integrated circuits could be realized in actual equipment. TI's 1961 contract also called for the construction of an IC pilot line within eighteen months that would be capable of turning out 500 integrated circuits per day for ten days. This requirement was clearly designed to push the IC concept toward reality.

It was Fairchild Semiconductor that provided the revolutionary *planar* process that moved IC technology from the laboratory to the production line. However, because of management policy, Fairchild neither sought nor received support from the government in its early IC development work. From 1959 to 1960, IC efforts were performed with in-house funds. In many ways Fairchild's work paralleled the work then being pursued at TI and Westinghouse. Fairchild's management knew of the programs, which were well publicized at these two companies and fully realized that its IC efforts competed with these.[21] Fairchild, then, was driven by the same stimuli as the other two companies but chose to go its own way.

Although the two basic patents and key technological contributions that underlie IC technology in the United States were made by private companies without government support, these fundamental innovations were achieved because both companies sensed the needs of their various customers, present and hoped-for. These customers, however, were drawn mainly from the government via its military interests. Thus, although government influence helped create the landscape these companies viewed, it did not dictate the nature of the technological route to be taken. The need was articulated, the means to satisfy it was not. In short, breakthroughs were brought about by the in-house R&D efforts of those companies that responded to the *articulated* demand of the military.

Japanese Research Consortia

Although the U.S. government was the primary customer for the semiconductor industry in the early stage of IC technology, its influence on the market decreased significantly in the years that followed. In 1963, the share of the federal government was 35.5 percent, in 1970, 20.6 percent, in 1972, 11.9 percent and in 1973, 5.8 percent.[22]

As a technology shifts from the defense sector to the civilian sector, the development of manufacturing technology becomes more important because cost is a critical factor in the civilian sector. Furthermore, as the shift to civilian sector occurs, many companies in different industries become involved in bringing the new technology into the consumer-products market, while only a few selected, technological elite companies are involved in the defense sector. In other words, the policy agenda shifts to building a national manufacturing infrastructure.

In the case of IC technology, the Japanese government, particularly MITI, played a significant role in the creation of this infrastructure. In 1976, MITI orchestrated the establishment of the ERA for VLSI development. The association existed from 1976 to 1979 and spent a total of ¥73.7 billion, of which ¥29.1 billion was paid by the government on a project funding basis. Members of the association were Fujitsu, Hitachi, Mitsubishi, NEC, and Toshiba.

Although we originally developed the concept of demand articulation to analyze the development processes conducted by a single firm, the dynamic process of collective action by rival firms creates the functional equivalent of demand articulation in a single firm. We can call this *collective articulation of demand.*

The collective articulation of demand, therefore, should be viewed in and can be explained by the overall framework of industrial technological linkages. It can assist in creating a national technological infrastructure. Sometimes it results in establishing upstream technological linkages. Indeed, the association for VLSI development made possible demand articulation for manufacturing equipment and materials for chip making.

The five companies established a joint research laboratory within the association. The laboratory had about 100 researchers who were drawn from the companies and from Electro-Technical Laboratory (ETL), one of Japan's national research institutes. Approximately 20 percent of the research was carried out in this joint research laboratory; the remaining 80 percent was done by the individual companies in their own laboratories with an association steering committee as a coordinating body.

The research was divided into six areas: micromanufacturing technology, with an emphasis on the development of equipment; crystal technology, with an emphasis on the analysis of the crystallizing process of silicon; product-design technology; process technology; testing and evaluation technology; and devices technology. All the research related to micromanufacturing technology and crystal technology was carried out in the joint research laboratory, while research related to product-design technology was carried out in the laboratories of member companies. As far as the R&D related to process technology, testing and evaluation technology, and devices technology was concerned, only the portions basic and common to all the members were carried out in the joint research laboratory.[23]

A great deal of the research and development carried out in the joint laboratory was subcontracted to supplier companies that were not members of the association, e.g., camera manufacturers, silicon crystal suppliers, and printing companies. Usually, in joint research by rival firms in the same industry, success hinges on ensuring that the research is basic and of common interest to all the participants. Therefore, rather than focusing on the method of producing chips, the association centered its research efforts around developing a prototype for IC manufacturing equipment and analyzing a process for the crystallization of silicon, a basic material in chip production. No manufacturers of production equipment or chip materials were among the participants.

Figure 4-3 depicts the major actors involved in the Japanese development of VLSI and the technical linkages between them.[24] The specific activities of the association included the development of the stepper, a piece of equipment used to reduce the electronic circuit onto the silicon base optically. Therefore, one of the association's lithography laboratories contracted the research necessary for the development of the stepper to camera manufacturers that owned the lens technology. Thus, companies such as Nikon and Canon succeeded in developing the stepper.[25] At the same time, the research results of the silicon crystallization process were passed along to the Shin-Etsu Semiconductor Company and the Osaka Titanium Company, both of which came to comprise Japan's silicon production industry.

FIGURE 4-3
Upstream linkages in Japanese VLSI development

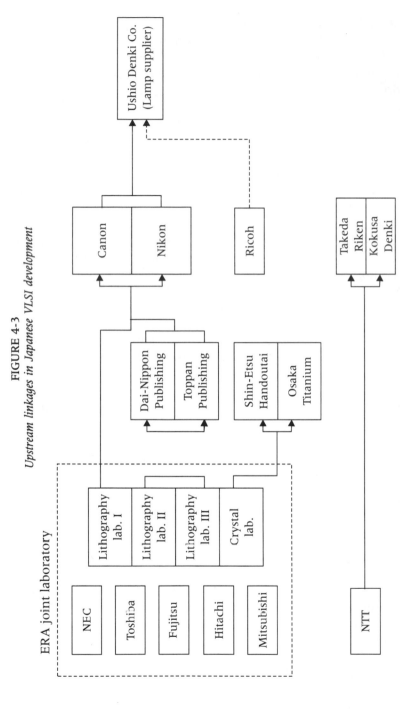

Source: J. Sigurdson, "Industry and State Partnership in Japan: The Very Large Scale Integrated Circuits Project," Research Policy Institute discussion paper no. 168 (Sweden: University of Lund, 1986), 86–93. Reprinted by permission.

TABLE 4-1
The world's top ten equipment manufacturers

	1980	1985	1989
1	Perkin Elmer	Perkin Elmer	*Tokyo Electron*
2	GCA	*Tokyo Electron*	*Nikon*
3	Applied Material	General Signal	Applied Material
4	Schlumberger	Varian	*Takeda Riken*
5	Varian	Teradyne	*Canon*
6	Teradyne	Eaton	General Signal
7	Eaton	Schlumberger	Varian
8	General Signal	*Takeda Riken*	*Hitachi*
9	Kulicke Soffa	Applied Material	Teradyne
10	*Takeda Riken*	GCA	ASM International

Source: V. Nicholas, "Thinking Ahead: A Survey of Japanese Technology," *The Economist* 313, no. 7631 (1989). 2–8.
Note: Japanese companies are in italics.

After ten years of demand articulation efforts, which were first initiated by the VLSI association, Japanese companies in the upstream sector of chip manufacturing are beginning to emerge as dominant players in world production. The changes in ranking of the world's top-ten suppliers of semiconductor manufacturing equipment are shown in Table 4-1.

Because we have said that collective demand articulation can create a national engineering infrastructure, we need to consider second-tier suppliers.[26] The suppliers of optical steppers, first-tier suppliers, were not the only beneficiaries of the joint effort. The real beneficiary was a second-tier supplier. Ushio Denki, the supplier of the lamp used for the optical stepper (see Figure 4-3), ended up dominating the world market for lamps. In 1983, Ushio had a market share of 100 percent for aligner lamps in Japan and 50 percent for the global market.

The Hierarchy and Multipolar Structure

A complex national system is necessary to articulate the demand for such a radical innovation as integrated circuit technology. Such a complex system is characterized by hierarchy and *multipolar* structure.

Although in the transition from the defense sector to the civilian market, leadership in IC technology shifted from the United States to Japan, the importance of demand articulation remained the same.

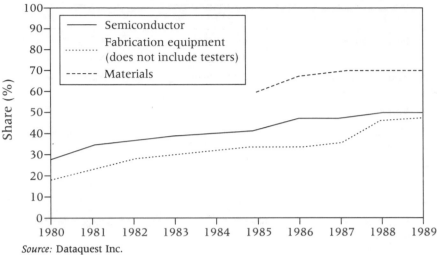

FIGURE 4-4

Changes in world market shares of Japanese semiconductor-related products

Source: Dataquest Inc.

Furthermore, the hierarchical nature of demand articulation could be clearly observed in both markets and countries. In the defense sector, the hierarchy was a *policy* hierarchy: the successful translation of the defense strategy into a technological concept contained a national security policy level, a system requirements level, and a component technological level. In civilian sector, the hierarchy was *manufacturing* hierarchy: the articulation of the demand for equipment and materials for chip making involved the successive translation of the demand from chip manufacturers to first-tier suppliers, and from first-tier suppliers to second-tier subcomponent suppliers.

Figure 4-4 shows how the Japanese share of the world production of semiconductors, fabrication equipment, and materials has grown. Although the country's market share of semiconductor production was 51 percent of the world total in 1989, it was 49 percent in fabrication equipment (excluding testers). As seen in the figure, the market share of materials production had surpassed that of semiconductors earlier and was as high as 71 percent in 1989.

A closer look at the changes since 1980 reveals that the production share of fabrication equipment caught up quickly, but it did not reach that of semiconductors until after 1987. It is not easy for equipment suppliers to penetrate foreign markets as they have to provide good technical service to their users.[27] Once we have taken this factor into account, we can say that the international competitiveness of Japanese fabrication equipment has surpassed that of semiconductors. This suggests that Japanese dominance in world semiconductor manufacturing becomes even more pronounced when it moves toward such upstream sectors as manufacturing machinery and materials.

So far we have described the hierarchical nature of demand articulation, but demand articulation is also multipolar when it involves public policy. The U.S. Air Force hoped that molecular electronics would eliminate unnecessary materials and consequently greatly simplify and improve the life and reliability of future electronic circuitry. The molecular electronics concept (the knowledge at a microscopic level that would allow proper utilization of every essential atom and molecule) per se, however, proved quite controversial, and the air force came to support two parallel and competing programs: a much more ambitious and risky effort with Westinghouse and a less risky program with Texas Instruments. In the spring of 1959, it awarded a $2 million research contract to Westinghouse, but later in the year, as we have already noted, it awarded a two-and-a-half-year contract to TI. In 1961, Westinghouse exhibited a radio receiver demonstrating its molecular electronic principle. As a digitial system, however, the TI equipment was a better example of IC technology.

Although the TI project had been initiated as an in-house program without government support, Jack Kilby, the inventor of the integrated circuit, had worked at Centralab, a company that had pursued miniaturizing electronic components in earlier work with the National Bureau of Standards (NBS) before joining TI. Centralab's "Project Tinkertoy," which was sponsored by the U.S. Navy's Bureau of Aeronautics, was perhaps the first major miniaturization effort. The work, however, was performed under the auspices of NBS. Furthermore, a key staff member who contributed to Kilby's work at TI had pioneered in the use of photolithographic techniques for semiconductor devices at the army's Diamond Ordnance Fuze Laboratories (DOFL). In 1957, DOFL had begun to work on ways to fabricate smaller electronic assemblies via a two-dimensional construction.[28]

MITI's project articulated the demand for optical steppers and materials, but an NTT project articulated the demand for testers. The VLSI research association paid little attention to the importance testers, but from 1977 to 1981, NTT conducted joint research on the next generation of testers with the Takeda Riken Company, which then became the major supplier of memory testers. In this joint research, detailed requirements for the new tester were collected from NEC, Fujitsu, Hitachi, and Oki, the major VLSI manufacturers. However, the fundamental requirements were eventually set by NTT after several meetings to work out the joint specifications (see Figure 4-3).

MODELING THE PIPELINE VIEW

So far we have contrasted the process and structure of demand articulation with the pipeline view of product development. These, however, are only mental models of product innovation. By using two types of collective research, we can transform these two contrasting views of product development into empirical studies, which may make the distinction between them even more explicit.

To do so, we need a classification scheme for collective research. One obvious taxonomic question to ask is whether the collective research is centered around the concept of public goods, or of private goods. If collective research is organized to gain common basic knowledge, i.e., based on the public goods concept, we can assume that participation in a project mirrors the participants' pipeline view of product development. If, however, collective research is organized to acquire and to develop technical know-how, which can be commercialized by participating firms, i.e., based on the private goods concept, we can assume that participation in a project reflects the demand articulation view.

Now that we identified two categories of collective research, we need to identify two specific systems of research collaboration that reflect the distinctions between these categories, and we need to analyze participation in these two systems of collective research.

Database

Usually international collaboration is based on the public goods concept of technology. To date, the majority of international collaborations, with a few exceptions, have been scientific projects, such as international collaborative Antarctic expeditions and global climate research projects. These collaborations aim at gaining common, basic knowledge. Thus, they are organized and managed according to the pipeline view.

From several possibilities, we chose the international joint projects sponsored by the International Energy Agency (IEA). The IEA was established in November 1974 at the recommendation of the Organization for Economic Cooperation and Development (OECD) and its membership covers twenty-four countries. It is an autonomous institution within the OECD. Its objective is to develop both conventional and alternative energy technologies. Its major activity is to create and manage international collaboration in energy-related R&D projects. Projects require, among other things, the participation of at least three countries and a limited time frame. A project proposal is reviewed by the working party of the IEA, which nominates a lead country. The proposal is then distributed to all the member countries to ask for participation. Each country decides on its own level of participation without consulting others.

In 1984 there were sixty-one projects[29] covering the following subjects (the number of projects so far implemented in each subject category in parentheses):

Systems analysis	(1)
Energy use	(24)
Fossil fuels	(9)
Geothermal energy	(2)
Solar energy	(9)
Bio-mass energy	(2)
Ocean energy	(1)
Wind energy	(6)
Hydrogen energy	(3)
Nuclear fusion	(4)

The systems analysis project was IEA's first project and very general. It also has the most participants—as many as eighteen countries. Because it is so different from the other projects in its nature and size, we excluded it from the sample.

Because participation varies from country to country, for example, the United States and Sweden have participated in 70 percent of all the projects while Mexico and Finland have participated in only one project, we selected for the final database only those countries that have participated in more than two projects as a lead country. There were ten countries that satisfied this condition.

Independent Decision

If members of the IEA assume the pipeline view, each will decide independently whether or not to participate in a project. We can deduce intuitively that this independent decision making will produce a bell-shaped frequency distribution of the number of participants: a certain number of participants, somewhere between one and all possi-

TABLE 4-2
Frequency of number of participants in an IEA project

Number of Participants	Observed Frequency
1	1
2	2
3	11
4	13
5	10
6	10
7	4
8	6
9	2
10	1
TOTAL	60

ble participants, can be expected, with a maximum probability of participation around the average.

The frequency distribution of the number of the ten countries that participated in the sixty projects is displayed in Table 4-2. Figure 4-5 depicts the frequency distribution curve, which is bell-shaped. Because each country makes its decision independently, we can think of two types of models: static and dynamic.

Static Modeling. Let p be the probability that a country participates in a project, then the probability distribution of the random variable X (the number of participating countries) follows a binomial distribution expressed mathematically by

$$Pr(X = k) = {}_nC_k \cdot p^k \cdot (1 - p)^{n-k},$$

where k is the number of participants, and n is the total possible number of countries.

By letting the participating probability, p, be such that the observed average coincides with the calculated average, $n \cdot p = 5.07$, and by letting $n = 10$, the probability distribution can be estimated. The difference between the observed value and the calculated value is shown in Table 4-3. It looks as if a binomial distribution could explain the observed values. The χ-square value, however, is 26.38, a little bit above 18.31, the 5 percent point for a χ-square distribution with 10 degrees

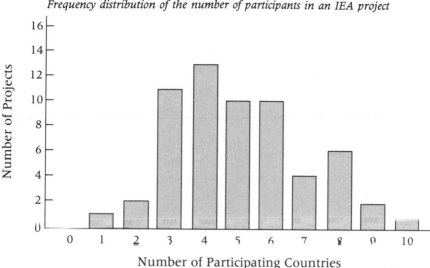

FIGURE 4-5
Frequency distribution of the number of participants in an IEA project

Number of Participating Countries

of freedom. Therefore, using static modeling, we cannot prove statistically that decision making is independent.

Dynamic Modeling. There are various ways to modify the theoretical models in order to improve the fit. One method is to formulate the participation process in an analogy of *random arrival.*

A distribution of the number of random arrivals during a limited time-interval produces the Poisson distribution when arrival phenomena are stochastically independent of each other. Therefore, it is possible for us to build a model in which the random arrival of countries in a specified time-period is equivalent to the participation of countries in a project, when the decision to participate is made independently.

Let μ be the average number of arrivals in a given interval, then the probability that the number of arrivals is k, $Pr(k)$, is described by

$$Pr(k) = (\mu^k/k!) \cdot e^{-\mu}.$$

By letting μ be the observed average, $\mu = 5.07$, the probability distribution can be estimated. The difference between the observed value and the calculated value is shown in Figure 4-6. The χ-square value is 6.30, far below 18.31, the 5 percent point for χ-square distribution with 10 degrees of freedom.

After studying the figure, we can say that the bell-shaped frequency distribution curve of the number of participants reflects very indirectly

TABLE 4-3
Difference between observation and binomial model

Number of Participants	Observed Frequency	Binomial Model
0	0	0.05
1	1	0.53
2	2	2.37
3	11	6.66
4	13	11.97
5	10	14.75
6	10	12.63
7	4	7.41
8	6	2.85
9	2	0.01
10	1	0.00
χ-square value		26.38

FIGURE 4-6
Difference between Poisson model and observed frequency

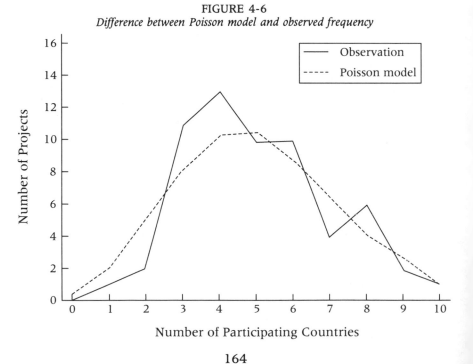

164

their view that the research outputs of collaborative projects are public goods. Thus, each participant holds the pipeline view of product development, not the demand articulation view. In this type of collective research project, then, our intuitive deduction is valid—a certain number of participants somewhere between one participant and all possible participants are to be expected, with a maximum probability of participation around the average.

MODELING THE DEMAND ARTICULATION VIEW

In the same manner in which the pipeline view was brought into an empirical study of IEA-led projects, we will try to bring the demand articulation view into an empirical study of research association projects. Most collective research projects conducted within national borders are based on the private goods concept of technology.

Database

The most appropriate database for this study would be a research collaboration system in which only rival firms participate. At the same time, however, the sample must be large enough for statistical inference.

In 1961, the Japanese government enacted the Engineering Research Association Act. Under this law, any collective research activity designated by the government as an ERA can receive the following tax benefits: (1) *tax deduction* of expenditures charged by cooperatives to their members for the membership fee and the acquisition of fixed assets for research; (2) *condensed recording*, as small as one yen, of charges imposed by cooperatives for the acquisition of fixed assets required for the study of technology; and, (3) *accelerated depreciation* of two-thirds of fixed assets for research in the first three years.

Since 1961, more than 100 ERAs have been established, and quite a number of them have completed their missions and have been disbanded.[29] The frequency distribution of the fifty ERAs that existed in 1985 is shown in Figure 4-7. The number of participants is distributed in a wide range from three to as many as forty-two (Technology Research Association of Medical and Welfare Apparatus). There is no regularity in this distribution curve. One could possibly say that it follows approximately a bell-shaped normal distribution, but, all that we can say with certainty is that this distribution occurs because of the

FIGURE 4-7

Frequency distribution of the number of participating firms in ERAs active in 1985

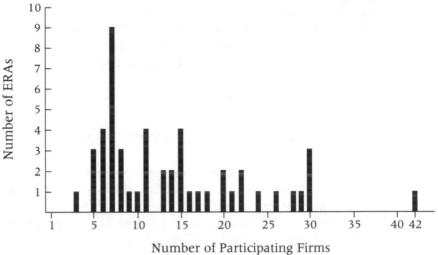

Number of Participating Firms

central limit theorem: if the samples are drawn from several populations of different categories, they distribute normally.

Rival Firms Participation. We are interested in collective action among competing firms. Therefore, only those associations composed of rival firms were selected. Those industrial sectors in which more than three companies of a similar size exist were chosen for study.[30] The industrial sectors included steel, textiles, computers, and shipbuilding, as shown in Table 4-4.

The frequency distribution of the number of participating firms is shown in Table 4-5.[31] As the table indicates, there were as many as sixty-seven cases of collective research in which any one of the com-

TABLE 4-4

Rival firms selected for analysis

Steel Companies	Textile Companies	Computer Companies	Shipbuilding Companies
Shin-Nihon	Toray	Toshiba	Mitsubishi
Nihon-Kohkan	Teijin	Hitachi	Kawasaki
Kawasaki	Asahi-Kasei	Fujitsu	Ishihari
Kobe Steel		Mitsubishi	
Sumitomo		NEC	

TABLE 4-5
Frequency distribution of participation by selected rival companies, 1988

Industrial Sector	Number of Projects	Number of Firms Participating				
		1	2	3	4	5
steel	40	27	1	1	4	7
textile	17	11	2	4	—	—
computer	67	27	15	6	4	15
shipbuilding	37	17	9	11	—	—

Source: I. Shirai and F. Kodama, "Quantitative Analysis on Structure of Collective R&D in Japan," NISTEP Report No. 5 (Tokyo: National Institute of Science and Technology Policy, 1989).

puter manufacturers participated, forty cases in the steel industry, thirty seven in the shipbuilding industry, and seventeen in the textile industry. The distribution curve of the number of companies participating indicates that the pattern of participation is U-shaped in every industrial sector, almost opposite of the pattern observed in Figure 4-5. This U-shaped pattern may look counterintuitive because it is the opposite of the bell-shaped pattern that could be deducted intuitively from the assumption of independent decision making on participation.

NTT-Led Joint Research. Although the pattern of rival firm participation in ERAs alludes to a U-shaped distribution, the sample is not large enough for statistical inference. Therefore, we need another source. NTT and communication equipment manufacturers have conducted a substantial number of joint research projects. NTT used to be a public company, and there are many similarities between the projects organized by NTT and the projects in an ERA. Because the telecommunication business can be inherently monopolistic, NTT is prohibited by law from manufacturing. Therefore, its business configuration is that of service industry and similar to public service. Furthermore, joint research organized by NTT is among rival firms, as is often the case in an ERA.

Although the number of joint research projects organized by NTT is not known, we can estimate it by counting the joint applications for patents.[32] Although we cannot always assume that a joint patent application is the result of joint research, it is hard to imagine joint patent applications not having joint research and single applications being the result of joint research. Even so, we cannot assume a one-to-one correspondence between a joint application and joint research, because one joint research project might produce several patent appli-

TABLE 4-6
Number of joint patent applications

Year	Total Number	Number of Manufacturers Involved			
		1	2	3	4
1980	563	457	50	7	49
1981	669	426	59	20	164
1982	652	479	33	8	132
1983	655	487	49	25	94
1984	540	374	28	5	133
1985	375	270	18	15	72
1986	655	532	53	22	48

cations. However, we are concerned with the participation pattern in joint research; therefore, we can safely assume that the statistics of joint application reflect the participation pattern of joint research.

In 1986, NTT with one or more of the four equipment manufacturers applied for as many as 655 patents. The time-series data of the joint applications is shown in Table 4-6. As we can see from the table, NTT cooperated most frequently with only one equipment manufacturer. Cooperation with all four manufacturers was second most frequent, and cooperation with two or three manufacturers was less frequent. Although the number of cases involved is substantial, this pattern of participation is quite stable over the seven-year period.

In order to avoid yearly fluctuations, however, we calculated three-year moving averages, and used them for our analysis. The relative frequency-distribution curve is depicted in Figure 4-8. Roughly speaking, cooperation with only one manufacturer accounted for 60 to 70 percent of the applications; cooperation with all four manufacturers accounted for 15 to 20 percent; and cooperation with two manufacturers and with three manufacturers accounted for 8 to 10 percent and 2 to 3 percent, respectively. This is an obvious manifestation of U-shaped distribution.

Interdependent Decision

Now we know that the distribution of the number of participating rival firms in collective research has a U-shape. We will analyze this somewhat counterintuitive phenomenon using mathematical models

FIGURE 4-8

Frequency distribution of number of equipment manufacturers involved in joint patent applications with NTT (three-year moving average)

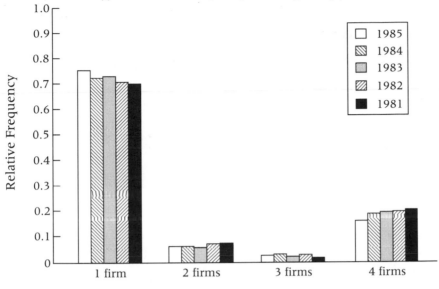

Number of Firms Involved

of interdependent decision making. We can argue that this interdependent decision making is the result of the demand articulation view of product development. Through the articulation of demand, participants expect to acquire a better identification of product specifications and thus a better cost estimation. Therefore, a firm's decision to participate is based on its estimation of benefit and cost. This type of decision is interdependent because the benefit each participant will receive and the cost each will have to pay depend upon the other participants.

In this context, decisions to participate can be formulated within the framework of a *noncooperative game* in which the four equipment manufacturers are players.[33] The rules of the game are as follows: (1) NTT asks the four players *one by one* about their intention to participate in the game. The game is terminated after NTT asks all the four players. (2) Each player is informed of the benefit and the cost of all the possible combinations of participation.

Each player then behaves so as to maximize payoff, i.e., the benefit minus the cost. Game theory tells us that each player decides to participate only if the benefit exceeds the cost. Therefore, if we know the rule by which benefits and costs are allocated, we can formulate the decision structure of participation. Although there are many variations of allocation rules, let us assume the following simplest set of rules:

1. The total cost of the project is independent of the number of participants. Therefore, it is denoted by c.
2. The total benefit of the project is independent of the participation pattern and is denoted by b. It is allocated equally to the participating players.
3. The cost c is shared equally by NTT and participants.

Now, we can derive the payoff of the player. Let n be the number of participating players. The payoff $U(n)$ can be described as

$$U(n) = b/n - c/(n + 1).$$

Let $z = b/c$ (the benefit/cost ratio), then

$$U(n) = c \cdot [z/n - 1/(n + 1)].$$

therefore,

$$U(1) = c \cdot (z - 1/2),$$
$$U(2) = c \cdot (z/2 - 1/3),$$
$$U(3) = c \cdot (z/3 - 1/4),$$
$$U(4) = c \cdot (z/4 - 1/5).$$

Each player will decide to participate if the payoff is positive and will decide not to participate if the payoff is negative. Therefore, the participation condition can be derived as follows:

if $0 < z < 1/2$,   then *no firm* participates;

if $1/2 < z < 2/3$,   then *one firm* participates;

if $2/3 < z < 3/4$,   then *two firms* participate;

if $3/4 < z < 4/5$,   then *three firms* participate;

if $4/5 < z$,   then all the *four firms* participate.

If we follow the rule of equal cost-sharing, there is no participation when z is less than 1/2. If NTT is ready to share more than half the

FIGURE 4-9
Relation between benefit/cost ratio and cost shared by NTT

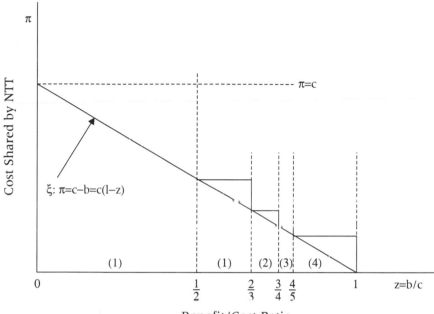

Benefit/Cost Ratio

cost, however, it is possible for one firm to participate. If NTT shares more than $\pi = (c - b)$, the payoff becomes positive for one firm. The alternative for NTT is to do it alone and carry the full cost. It is reasonable to assume that NTT is ready to share the cost of $\pi = (c - b)$, and therefore, even if z is less than $1/2$, we can assume that one firm will participate. Thus, if $0 < z < 2/3$, one firm will participate.

On the other hand, if z is larger than 1, the benefit exceeds the cost without any sharing. In other words, it pays off if any manufacturer, not just one of the four equipment manufacturers, develops the technology alone. In other words, NTT has no reason to initiate joint research because it will receive the benefits generated by the in-house research of the manufacturer. The same situation applies to any one of the four manufacturers. Therefore, we can assume that it is not possible for joint research to be realized. Thus, all the four firms will participate only if $4/5 < z < 1$.

Statistical Tests

The participation condition derived from this analysis has been integrated in Figure 4-9, where the benefit/cost ratio, z, is plotted along

TABLE 4-7
Difference between observation and values calculated by uniform distribution of z

Participation Pattern	Calculated	Observed
One-firm participation	66.7%	70.4%
Two-firm participation	8.3	7.1
Three-firm participation	5.0	2.7
Four-firm participation	20.0	19.7

Note: Observed are three-year moving averages at 1982.

the X-axis and the cost shared by NTT, π, is plotted along the Y-axis. As we can see in the figure, the number of participants is uniquely determined by the value z of benefit/cost ratio. Therefore, we can derive the probability distribution of the number of participating firms, if the probability of z is given.

There is no relevant information concerning the mathematical form of the probability density function of z. Therefore, we will tentatively select uniform distribution because it is the simplest mathematical form.[34] Distribution is then proportional to the length of intervals between the boundary values (see Figure 4-7). The results are shown in Table 4-7. It is obvious that this model regenerates the basic distribution characteristics of participating firms: one firm's participation is the most frequent, four firms' participation is the second most frequent, two firms' participation is the third and three firms' participation is the least frequent. Therefore, we can say that this model simulates the basic characteristics of the frequency distribution of participation behavior, at least to the first order. Thus we can conclude that participation is based on *dependent* decision making.

If a statistical significance test is applied, however, it yields the following χ-square values: 15.42 for 1981; 9.53 for 1982; 17.94 for 1983; 10.41 for 1984; and 26.09 for 1985. We cannot say that the model fits the observations statistically, because the 5 percent point is 7.78 with 3 degrees of freedom. However, there are several ways to improve the model's performance.

Generally, we can say that a collective action becomes necessary when the benefit/cost ratio is lower than the average. Therefore, it is reasonable to assume that the distribution of the benefit/cost ratio of z in collective research is skewed toward smaller values between 0 and 1. Therefore, we can modify the uniform distribution of z, to a decreasing function of z, for example, such as $f(z) = 1.3 - 0.6 \cdot z$. By using the

FIGURE 4-10
Difference between observation and decreasing function model

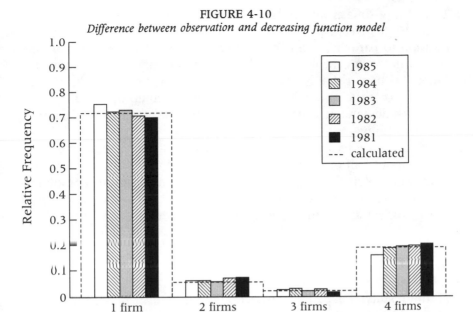

Number of Firms Involved

formula in the footnote, we can calculate the participating frequency distribution. Then, as shown in Figure 4-10, the difference between observed and calculated values becomes minimal. Needless to say, the statistical test of significance has no problem at all. A statistical test, for example, yields the following χ-square values:

4.88 for 1981;
2.57 for 1982;
1.03 for 1983;
0.62 for 1984; and
8.28 for 1985.

Let $f(z)$ be the probability density function of z, then the probability that the number of participants is n, $q(n)$, is derived by the following formula:

$$q(1) = \int_{z=1/2}^{z=2/3} f(z) \cdot dz + \int_{z=0}^{z=1/2} f(z) \cdot dz,$$

$$q(n) = \int_{z=n/(n+1)}^{z=(n+1)/(n+2)} f(z) \cdot dz, \quad \text{for } n = 2, \text{ or } 3,$$

$$q(4) = \int_{z=4/5}^{z=1} f(z) \cdot dz.$$

Thus, the χ-square values for all the years except 1985 are far below 7.78, the 5 percent point with 3 degrees of freedom. Therefore, the decision to participate in collective research is interdependent among rival firms. We can also conclude that participation is economically rational behavior that is based on estimates of benefit/cost ratios. This indicates that the rationale for collective research among rival firms is collective demand articulation, although the observed phenomenon is counterintuitive.

Because the two extreme cases are more plausible than the cases where about half of all possible candidates are to participate, intuitive judgments are misleading in organizing research consortia. The advantages of collective research among rival firms need to be viewed in a broad context and with a long-term perspective.

SOCIAL LEARNING

As the results of Japan's high technology began to be visible around the world, many observers began to view research associations as conspiracies between the government and the business community.[35] Our analysis of collective demand articulation, however, proves that theories of collusion or conspiracy are not needed to explain corporate cooperative research, even applied research or development.[36] Revisionists who claim that Japanese society is working under different principles from the principles of Western cultures[37] have failed to grasp the rationale behind collective research, but they are not alone. It took a long period of trial and error before Japanese government and industry discovered this rationale.

We can, in fact, argue that many of MITI's policy innovations have been an unconscious part of the effort to catch up with more advanced countries.[38] Certainly the development of the ERAs was not free of mistakes. Rather than trying to describe definitively how a research consortium should be managed, we can learn more by reviewing the establishment of ERAs in Japan.

Trial-and-Error Process

Britain first introduced the research association scheme in 1921. From Britain, the idea spread to such countries as France, Germany, Italy, and Norway. In 1961, Japan reformulated the idea into today's ERAs.

European research associations were intended to support and enhance the technological ability of small businesses that were lagging

behind in modernization. In contrast, the purpose of Japanese cooperative research has been to raise the technological expertise of major
firms to the world level and then to the top of that level.[39] Because
the Japanese associations aim for the most advanced technology possible, their research objectives are set a generation ahead of the existing
technology. Current differences in the technological levels of participating companies are, therefore, not a great problem, and conflicts of
interest do not arise.

Prior to the creation of ERAs, collective research was carried out in
voluntary organizations. Because voluntary associations did not have
the status of *legal* corporations, no tax incentive could be derived from
collective research even if it produced social benefits. In order to provide an incentive for collective research that may yield social benefits,
the Engineering Research Association Act was enacted in 1961. As
noted earlier, under this law, a collective research association designated by the government as an ERA can receive several tax benefits.
In October 1961, the first ERA was established to conduct research
in polymer materials technology. Throughout the 1960s, ERAs were
established to bring Japanese technology up to the world level. In the
mid-1970s, the emphasis shifted from catching up to generating basic
technologies. The focus moved to long-term national research programs. The epoch-making example of this shift was the ERA for VLSI
development, established in March 1976.

Between 1961 and 1985 a total of seventy-six ERAs had been established, fifty of which were active as of July 1985. In other words, by
1985, twenty-six ERAs had completed their missions and had been
disbanded.[40] The time trend in the number of ERAs established in each
year is shown in Figure 4-11.

In 1966, MITI launched the National R&D Program, popularly
known as the large scale project (LSP). This was the first attempt by
the Japanese government to finance 100 percent of the R&D carried
out by private enterprises. The defense R&D programs in the United
States and Europe were the government's models for this project. MITI
assumed that the LSP was incompatible with the ERA scheme. Therefore, as shown in Table 4-8, no new ERAs were established in the four
years following the implementation of the LSP system, although as
many as five large scale projects were launched during this period.

In 1977, during an R&D project on a pattern information-processing
system (PIPS), MITI discovered that an LSP was compatible with the
ERA and that, in fact, they complemented each other. It has since
become normal practice to use ERAs for the implementation of MITI's
R&D programs.

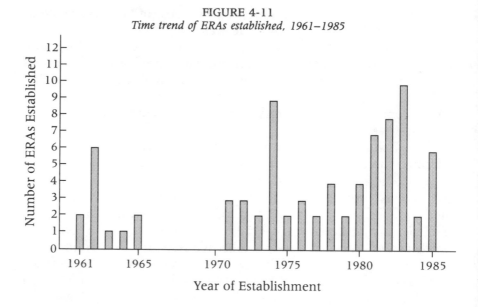

FIGURE 4-11
Time trend of ERAs established, 1961–1985

Year of Establishment

PIPS Project

PIPS was launched under the guidance of ETL. As many as ten companies participated in this project and as many as eighteen research items were investigated. ETL was responsible at least for some parts of all of them. PIPS continued until 1980, when total funding reached ¥22 billion. There were four areas of research: (1) pattern recognition systems; (2) devices and materials; (3) information-processing systems; and (4) an integrated system prototype.[41]

Those research items related to research areas (1) through (3) were contracted directly to the ten companies, while those related to (4) were contracted to five companies: Hitachi, Toshiba, NEC, Fujitsu, and Mitsubishi Electric. The integrated system has the following six recognition subsystems: printed *kanji;* handwritten characters; gray picture; color picture; object; and speech. Each firm was responsible for one subsystem.

Because PIPS was an LSP, the research on each subsystem was carried out rather independently, and thus the subsystems were not as integrated as had been expected. In order to conduct research on an integrated system prototype, an ERA was established in 1977. This ERA came into existence as a voluntary organization in 1975, five years after PIPS was launched, but it became a legal corporation in 1977. It acted as a coordinating body for companies with differing

TABLE 4-8
Compatibility between ERA and large scale projects

Year	New Engineering Research Associations	New Large Scale Projects
1965	2	—
1966	0	2
1967	0	1
1968	0	0
1969	0	1
1970	0	1
1971	3	3
1972	3	0
1973	2	3
1974	9	0

hardware architectures, thus permitting them to exchange information and integrate a prototype system.

There seem to be two reasons why the LSP and ERA schemes are compatible: one is administrative and the other is technological. When the government directly contracts firms, all the detailed administrative work must be performed within the government, which is not well prepared organizationally or technically for this kind of work. In order to perform the work smoothly, the government has to hire extra staff who become government employees with rigid job assignments and an inflexible bureaucracy. It is better if the staff is employed by the ERA, which has a more flexible bureaucracy than the government.

In terms of technology, the two schemes are compatible when it is an R&D project involving systems integration. In this type of project, the government assigns contracts for various component technologies to different companies. Therefore, some kind of collaborative effort is needed when the system's comprehensive design stage is reached and during the final testing and evaluation of the integrated system. However, neither the government nor the participating firms are the appropriate vehicles to coordinate the effort; the ERA is.

Learning by the Industry

After much trial-and-error, Japanese industry has come to appreciate the research association scheme. A survey conducted by Kei-Dan-Ren

TABLE 4-9
Industry's evaluation of government R&D policies

Policy Category	Evaluation Score (full-score = 100)
Budget size	56
Human resources	57
Research facilities	59
Government-industry collaboration	61
International relations	62
Promotion of basic research	64
Intellectual property rights	67
Standardization	68
Policy organization	68
Dissemination of information	70
Subsidy	65
Loan and investment	66
Research contract	68
Tax incentive	71
Research association	72
AVERAGE	64

Source: The Federation of Economic Organization, "The Evaluation Survey on R&D Policies," Economic Report No. 356 (in Japanese) (Tokyo: Kei-dan-ren, 1989).

(The Federation of Economic Organization) asked member companies to evaluate various government R&D policies (see Table 4-9). Among the fifteen policy categories, research association received a score higher than the scores given to such categories as subsidy, loan and investment, and research contract. This would seem to indicate that managers are not uneasy about government intervention when it occurs through government-sponsored research consortia.

One of the major issues in any government's industrial technology policy is how to manage the science-technology interaction, particularly the interaction between national research laboratories and industry. An ERA can be a channel for transforming scientific ideas into technical innovations.

In many cases, the research director of an ERA is recruited from a national research laboratory because he or she has the most advanced knowledge and supposedly can be neutral about the vested interests of the participating firms. Neutrality is critical when conflicts of interest occur at the stage of integration, which happened in the PIPS project. Moreover, researchers at national research laboratories are heavily involved in planning work for the association and can make substantial contributions to setting project goals. At one time, during the PIPS project, the participating firms were reluctant to conduct research because they thought the goals set for each subsystem were not feasible. They were persuaded to do so because the ETL researchers had already conducted some basic research, which indicated the feasibility of the goals.[42]

INTERNATIONAL IMPLICATIONS

The collapse of communism and the resultant end of the cold war are making it clear that technology drives the world's political and economic order. Even so, international business has not succeeded in capitalizing on the newly emerging techno-paradigm. Only with a deeper understanding of the product development process can we think of various types of cooperation from which the world can profit within the competitive world-trading environment.[43]

Based on what we have described so far in this chapter, it should be clear that the incorporation of the principle of demand articulation into a company's existing technology strategy is the key factor for successful product development.[44] The incorporation of this process is a long-term process of change, but it is necessary if companies are to survive. Given this, management's first task is to revisit the relationship between the customer and research and development. By doing so, a high-tech firm can find the best way to capitalize on the demand articulation capabilities built into its customer industries, even when these industries are located in other countries. Thus, both parties can play a plus-sum game. There are at least two examples of this type of cooperation: one is a case in which U.S. material suppliers capitalized on the demand articulation made by a Japanese automobile company; the other is a case in which a Japanese electronic component supplier capitalized on the demand articulation made by an American computer company.

Toyota-3M Collaboration

Closed subcontracting relationships between large and small companies, often described as *supplier-Keiretsu*, are infrequent in the Japanese chemical industry.[45] Some international chemical companies, however, have been capitalizing on the demand articulation capabilities that are built into Japanese fabrication industries, such as the auto industry and the electronics industry. Minnesota Mining & Mfg. Co. (3M), for example, receives approximately 50 percent of its $13 billion in annual sales from outside the United States. The company has more than doubled its R&D investment in support of its international business over the past five or six years. In 1960, the company's slogan was "Research Is the Key To Tomorrow." Today, its slogan is "Innovation Working for You," which implies that the company's R&D efforts, as well as its marketing, manufacturing and other functions, have become focused on the customer. Moreover, what truly differentiates 3M from other chemical companies is *not* its number of technologies, but rather the unique and innovative ways they are combined to solve customer problems.

3M enters market segments they can win either because the market is growing, the market is prepared for a change, or the company can change it through innovation.[46] For example, 3M had developed a core technology in acrylic adhesive foam tapes, which the auto industry uses to attach body side moldings. The base technology was developed in 3M's U.S. labs, but, when it was transferred to Germany, costs and customer value concerns led to the need for further development. The 3M labs in Germany did this, and because of the excellent communication and technology transfer in the 3M technical family, the company's laboratories in Japan knew of the developments. However, 3M's customer in Japan—Toyota Motor Corp. in particular—had some specific performance requirements for the use of the product in their Lexus models. To meet these requirements, 3M's Japanese laboratory advanced the technology to the next level. Now, the technology is widely available to all U.S. auto manufacturers.

Sun-Fujitsu Alliance

As technologically progressive industries mature, entry into these industries becomes more difficult. When it does occur, it is likely to involve the formation of coalitions. Many small U.S. entrepreneurial firms that generate new, commercially valuable technology do not have critical component manufacturing competencies within their boundaries. Thus, they either have to incur the expense of trying to build them, or they have to try to develop coalitions with companies that have these competencies. This is now occurring internationally, particularly in the computer and telecommunications industry.[47]

In fact, the computer industry is likely to see a proliferation of strategic partnerships in the R&D phases, which will gradually expand into global consortia when the technology has carved out a market niche. United States-Japan alliances are becoming increasingly important because Japanese companies dominate many leading-edge technologies, such as LCDs, memory chips, IC cards, thin packaging, and mass manufacturing. The alternative for U.S. computer companies that do not have the critical components is to capitalize on Japanese component strengths to expand their worldwide market share. By the same token, large Japanese component manufacturers can capitalize on the demand articulation capabilities of small U.S. entrepreneurial firms.

The alliance between Sun Microsystems and Fujitsu Limited is only one example that demonstrates how a Japanese chip manufacturer could capitalize on a specific demand articulated by a U.S. innovative firm. In his study on U.S.-Japan strategic alliances in the semiconductor industry, Okimoto has asserted that the Sun-Fujitsu strategic alliance has been crucial to Sun Microsystems' success in the highly competitive workstation market.[48] Without Fujitsu's early commitment to produce 32-bit reduced instruction set computer (RISC) processors based on Sun's scalable processor architecture (SPARC) operating system, Sun might not have distinguished itself as an industry leader. The alliance has benefitted both companies. Sun, for example, has become the leader in the fast-growing workstation industry. The company achieved a half-billion dollars in sales in just five years and has captured nearly 25 percent of the Japanese workstation market.

Ironically, Fujitsu was not Sun's first choice for a partner. Sun approached most of America's large semiconductor companies about a possible partnership in SPARC architecture, but for a variety of reasons, none was willing or able to develop the SPARC chip. Sun needed a single chip (exclusive of floating point) for its Sun-4 workstation. In 1983, the year of Sun's search, a 20,000-gate array (semi-custom chip) with 4,906 bits of register file memory in static random access, which was needed for the task, was not commercially available from any U.S. chip maker. LSI Logic was selling 10,000-gate arrays and had designed a 50,000-gate array, but it would not ship silicon until 1986. Advanced Micro Devices, Fairchild, and Texas Instruments had technical solutions, but they were six to twelve months behind Sun's schedule. Fairchild had a 5-MIPS Clipper processor, but it was too slow. At the time, Sun was an unknown start-up company, and RISC was an unproven commercial technology, although it was attracting attention. Thus, microprocessor vendors preferred to focus on their own complex instruction set computer (CISC) microprocessors, and U.S. chip makers such as Intel, Motorola, and National Semiconductor tried to steer Sun to their standard microprocessor lines.

In a last-ditch effort, Sun went to Fujitsu Microelectronics, Inc. headquarters in Silicon Valley. Sun's timing was fortuitous since Fujitsu was seeking to invest in emerging technologies and growth companies. Producing SPARC chips for Sun, the company argued, could position Fujitsu in the high growth technical workstation market. Although it was not easy to convince Fujitsu's Tokyo management, which had a "mainframe mentality," Fujitsu's executive vice president and director, who had founded the company's semiconductor operations, liked Sun's proposal. Fujitsu already had a 20,000-gate array that was being considered by two mini-supercomputer start-ups, Convex Computer and Alliant Computer, that were already using Fujitsu's 8,000-gate arrays. In 1984, Fujitsu decided to commit engineers to Sun's task. Fujitsu began producing S-16, a 10-MIPS SPARC microprocessor for the Sun-4 workstation in 1985, and Sun workstations became a hit in the marketplace.

The Sun-Fujitsu alliance has evolved over time. In 1988, Sun FMI and Wind River signed an agreement to accelerate the use of SPARC in real-time computing markets. Fujitsu also signed a five-year agreement to market Sun workstations in Japan, which pushed Sun's workstations ahead of Sony Microsystems' NEWS workstations and Hewlett-Packard's machines. During this period, Sun had the opportunity to diversify its SPARC chip sourcing. Sun could have manufactured its own chips, but semiconductor manufacturing is a totally different business that would have required heavy up-front investments. Going alone would have been risky, especially because Sun had no semiconductor experience. Sun could have found a second source, but it decided to stay with Fujitsu.

The Sun-Fujitsu alliance has had a domino effect on computer companies and components suppliers worldwide. Mips Computer Systems Inc. has worked out a two-year exclusive agreement with NEC Corp. for its R3000 workstations. Hewlett-Packard Co. recently licensed its precision architecture to Hitachi Ltd., Mitsubishi Corp., and Samsung Co. Intel Japan KK has teamed up with six companies in the U.S. and Japan. Motorola Inc. will supply chips to IBM and Apple Computer Inc. for their workstations. Within the next several years, all major workstation makers will belong to a global RISC consortium.

The shift toward demand articulation in product development is changing the rules of competition in high-tech industries. It is fostering global alliances that are beyond the traditional technical alliances formed among domestic companies. Instead of one national champion competing with another national champion in another country, a global consortium, composed of manufacturers and suppliers located in various countries, is likely to compete with another global consortium.

CONCLUSION

The concept of demand articulation was introduced in order to find an accurate way of describing the dynamic process of product development. Demand articulation is a two-step process in which (1) as yet unarticulated wants and virtual markets are integrated into a product concept and (2) this product concept is decomposed into development agendas for its individual component technologies, including emerging technologies.

To quantify the demarcation between the pipeline view and the demand articulation of product development, mathematical models were made on participation behavior in two different kinds of collective research. In a collective research system based on the pipeline view, we find that the probability distribution of the number of participants in any single project is bell shaped. In a collective research system organized by the principle of demand articulation, on the other hand, we observe a U-shaped distribution of the number of participants in any one project.

This U-shaped phenomena arises from economically rational behavior by rival firms, in which the decision to participate is made on the basis of each firm's estimate of the benefit/cost ratio. In other words, the dynamic process of collective action by rival firms creates the functional equivalent of demand articulation in a single firm, which can be conceptualized as collective articulation of demand.

By reviewing the development of the integrated circuit, which began in the U.S. defense sector and was furthered in Japanese government-sponsored research consortia, we arrived at the concept of a national system of demand articulation. This concept is more effective for analyzing national technology policy than is the often-discussed national innovation system.

Finally, demand articulation was used to analyze how businesses could profit through international cooperation within the competitive world-trading environment. In other words, we demonstrated how a high-tech firm can find the best way to capitalize on the demand articulation capabilities of its customer companies when these companies are located in other countries.

This chapter has shown how demand articulation works at various levels: at the individual company level (home-use VTR), at the industry level (the ERA for VLSI), at the national level (U.S. Department of Defense IC development), and at the international level (3M and Toyota, Sun and Fujitsu).

Still, we need to ask if demand articulation will work at the global scale in dealing with such environmental problems as global warming or within national boundaries for problems relating to community and urban development. To discover the answer to these questions, the concept of demand articulation must be investigated further. There are at least three areas that need to be addressed.

First, is demand articulation necessary and/or effective in the development of technology for global environmental protection and urban development? The structure of demand articulation in these areas may be quite different from those in the areas described in this chapter. Who is the beneficiary of the development of those technologies? Who pays for the cost of development? Is there any mismatch between those two agents?

Second, is the process of demand articulation different when it is applied at global and community levels? Even if demand articulation is effective for the development of technology, the decision-making process for developing and adopting new technologies is different. For social and urban development, decision makers include citizens, developers, and local governments. For global environmental problems, national governments, worldwide regional offices, and international organizations are decision makers.

Third, who is the demand articulation agent at these levels? It might be local government in urban development and an international organization in global environment. Those organizations, however, do not have the technological capability for demand articulation. What is important, therefore, is how they can mobilize and organize the competencies needed for demand articulation.

There may be other questions to research, but these are challenging research subjects in technology management. Indeed, they are so challenging that the research community of technology management may have to be extended to include researchers from sociology, anthropology, and international relations. In fact, how to best organize these researchers is an interesting research subject in itself.

NOTES

1. R. Gomory and R. W. Schmitt, "Science and Product," *Science* 240 (1988): 1131.

2. J. Alic et al., *Beyond Spinoff* (Boston: Harvard Business School Press, 1992), 19.

3. R. Gomory, "From the 'Ladder of Science' to the Product Development Cycle," *Harvard Business Review* 67, no. 6 (1989): 99–105; Gomory and Schmitt, "Science and Product," 1131.

4. R. Rowthwell et al., "SAPPHO updated—SAPPHO phase II," *Research Policy* 3, no. 3 (1974): 258–91.

5. R. Gomory, *"From the 'Ladder of Science,'"* 99.

6. David Warsh, "A Japanese Deming for the United States," *Boston Globe,* 10 May, 1992.

7. R. Nelson, ed., *National Innovation Systems: A Comparative Analysis* (New York: Oxford University Press, 1993),

8. F. Kodama, "Rivals Participating in Collective Research: Its Economic and Technological Rationale," in *Science and Technology Policy Research: What Should be Done? What Can be Done? The Proceedings of the First NISTEP Conference on Science and Technology Policy Research,* ed. H. Inose, M. Kawasaki, and F. Kodama (Tokyo: Mita Press, 1991), 141–64.

9. N. Rosenberg, *Perspectives on Technology* (Cambridge: Cambridge University Press, 1976), 108–25.

10. E. von Hippel, *The Sources of Innovation* (New York: Oxford University Press, 1988).

11. S. Kline and N. Rosenberg, "An Overview of Innovation," in *The Positive Sum Strategy,* ed. R. Landau and N. Rosenberg (Washington, D.C.: National Academy Press, 1986), 275–305.

12. R. Nelson and S. Winter, *An Evolutionary Theory of Economic Change* (Cambridge, Mass.: Harvard University Press, Belknap Press, 1982), 256.

13. According to Webster's dictionary, articulate comes from the Latin *articulare.*

14. F. Kodama, "Demand Articulation: Targeted Technology Development," in *Technik-Politik: Angeschichts der Umwelt-Katastrophe,* ed. H. Krupp (Heidelberg: Physica-Verlag, 1990), 273–94. See also Japan Industrial Policy Research Institute, *A Study of Innovation Process* (in Japanese) (Tokyo: Sangyou-Kenkyushyo, 1989), 48–57.

15. N. Sawazaki et al., "A New Video-tape Recording System," *Journal of the Society of Motion Picture and Television Engineers* 69 (1960): 868–71.

16. The Vladimir K. Zworykin Award of the IEEE (Institute of Electrical and Electronics Engineers) was given to the inventor, Dr. Sawazaki, in 1986.

17. Organization for Economic Cooperation and Development (OECD), "Case Study of Electronics with Particular Reference to the Semiconductor Industry" (joint working paper of the Committee for Scientific and Technological Policy and the Industry Committee on Technology and the Structural Adaptation of Industry, Paris, 1977), 133–63.

18. K. Oshima and F. Kodama, "Japanese Experiences in Collective Industrial Activity: An Analysis of Engineering Research Associations," in *Technical Cooperation and International Competitiveness*, ed. H. Fusfeld and R. Nelson (Troy, N.Y.: Rensselaer Polytechnic Institute, 1988), 93–103.

19. OECD, "Case Study of Electronics with Particular Reference to the Semiconductor Industry."

20. Ibid.

21. Ibid.

22. Douglas W. Webbink, *The Semiconductor Industry: A Survey of Structure, Conduct and Performance*, staff report to the Federal Trade Commission (Washington, D.C.: U.S. Government Printing Office, 1977), 69.

23. Y. Tarui, "The Records of the VLSI Joint Laboratory" (in Japanese), *Shizen* (September 1981): 34.

24. J. Sigurdson, "Industry and State Partnership in Japan: The Very Large Scale Integrated Circuits Project," Research Policy Institute discussion paper no. 168 (Sweden: University of Lund, 1986), 86–93.

25. Ibid.

26. Ibid.

27. Ibid.

28. This documentation about IC development in the U.S. is confirmed by Dr. Jack Kilby: "I have reviewed the portions [of the manuscript for this book] with which I am familiar and found it to be completely accurate." (Personal letter to the author.)

29. Ministry of International Trade and Industry, *Summary of International Research Cooperation* (in Japanese) (Tokyo: Agency of Industrial Science and Technology, MITI, 1984).

30. I. Shirai and F. Kodama, "Quantitative Analysis on Structure of Collective R&D by Private Corporations in Japan," NISTEP Report No. 5 (Tokyo: National Institute of Science and Technology Policy, 1989).

31. Ibid.

32. In order to expand the database, collective research organized by the Japan Key Technology Center, which was established in 1985 as an extended version of ERA, are included.

33. M. Kobayashi, "A Mathematical Model of Collective Research" (in Japanese) (master's thesis, Saitama University, 1987).

34. A. Spence, "Cost Reduction, Competition and Industrial Performance," *Econometrica* 52 (1984): 101–21.

35. M. Kobayashi, "A Mathematical Model of Collective Research."

36. L. Fisher, "Need for High-Tech Consortia Stressed," *New York Times,* 12 January, 1989.

37. Ibid.

38. C. Prestowitz, "Japanese vs. Western Economics: Why Each Side Is a Mystery to Each Other," *Technology Review* (May–June 1988): 27–36.

39. F. Kodama, "Policy Innovation at MITI," *Japan Echo* 11, no. 2 (1984): 66–69.

40. K. Oshima and F. Kodama, "Japanese Experiences in Collective Industrial Activity."

41. Ibid.

42. Engineering Research Association of Pattern Information Processing System, *Demonstration Panel: Integrated System Prototype of Pattern Information Processing System* (Tokyo: ERA for Pattern Information Processing System, 1980).

43. F. Kodama, "Direct and Indirect Channels for Transforming Scientific Knowledge into Technical Innovations," in *Transforming Scientific Ideas into Innovations: Science Policies in the United States and Japan,* ed. B. Bartocha and S. Okamura (Tokyo: Japan Society for the Promotion of Science, 1985), 198–204.

44. F. Kodama, "The Fusion Game: Capitalizing on Japanese High Technology Development," *Harvard International Review* 15, no. 4 (1993): 16–19.

45. F. Kodama, "Technology Fusion and The New R&D," *Harvard Business Review* 70, no. 4 (1992): 70–78.

46. F. Kodama, *Nikkei Weekly*, 19 October 1992.

47. R. Mitsch, "R&D at 3M: Continuing to Play a Big Role," *Research-Technology Management* 35, no. 5 (1992): 22–26.

48. D. Teece, "Profiting from Technological Innovation: Implications for Integration, Collaboration, Licensing and Public Policy," *Research Policy* 15, no. 6 (1986): 285–305.

49. D. Okimoto et al., *U.S.-Japan Strategic Alliances in the Semiconductor Industry: Technology Transfer, Competition and Public Policy* (Washington, D.C.: National Research Council, 1992).

5

Innovation Pattern
From Breakthrough to Technology Fusion

While no one argues that Japan has not been successful in designing products and production systems based on foreign technologies, critics often say that Japanese industries are adapters as much as innovators.[1] These critics frequently cite as proof the study made by Gellman Associates for the U.S. National Science Foundation.[2] The Gellman study found that between 1953 and 1973, Japanese scientists developed only 2 of 100 radical breakthroughs introduced into the marketplace. More recently, however, F. Narin and J. Frame have analyzed all U.S. patents issued between 1975 and 1985 in order to determine if the copycat characterization of Japanese technology is valid today.[3] They have concluded that patent statistics do not support this view.

Comparing the Gellman and Narin and Frame studies, we can see a tremendous gap between perception and reality. The gap is so large that it is hard to believe it originates solely from differences in methodology and/or time periods. According to R. Nelson, one should look at what has happened since the late 1960s not in terms of the degradation of U.S. performance, but rather in terms of the technological capability and management of other countries, primarily Japan but also Germany.[4] During this period, he argues, many other countries have caught up with the U.S. performance.

I believe there is an even more profound reason for what has happened: a drastic change in patterns of innovations. To understand this, we need to understand the intrinsic nature of technology. In his book *Inside the Black Box*, Nathan Rosenberg argues that descriptions of innovation pay little attention to the network of technological relationships in which specific inventions are embedded. He writes:

> The public image of technology has been decisively shaped by popular writers who have been mesmerized by the dramatic story of a small number of major inventions—steam engines, cotton gins, railroads, automobiles, penicillin, radios, computers, and so on. In addition, in the telling of the story, overwhelming emphasis is placed on the specific sequence of events leading up to the decisive actions of a *single individual* [emphasis added]. Indeed, not only our patent law but also our history textbooks and even our language all conspire in insuring that a single name and date are attached to each invention.[5]

Rosenberg then formulates the significant characteristics of technologies, their complementarities, their cumulative impacts and their interindustry relationships. Rosenberg finishes by noting that the inability to take these characteristics into account is the fundamental limitation of most recent literature on technological innovation. Technological progress in one sector of the economy has become increasingly dependent on technological change in other sectors. Such interindustry flows of technology are one of the most distinctive characteristics of advanced industrial societies today.

Certainly, Rosenberg's characteristics help to explain some things, but are they sufficient to explain the gap between the Gellman and the Narin and Frame studies? Rosenberg's case studies do not include civilian high technologies in which Japan has demonstrated her international competitiveness. Furthermore, we need to note the difference in the time periods of Gellman's study (1953 to 1973), and Narin and Frame's study (1975 to 1985). It seems quite clear that a techno-paradigm shift from modern technologies to high technologies, which we will describe later in this chapter, occurred in the mid-1970s. In this case, the Japanese experience may be creating a new type of innovation pattern.

COMPARING NATIONS' INNOVATIVENESS

According to Nelson, during the first twenty years after World War II, Americans became used to the idea that they were the world's technological and economic leaders. Nobody asked why they were or where that lead had come from. However, the lead started to erode in the late 1960s, and the erosion became conspicuous in the early 1970s.[6]

In the early 1970s, the rate of productivity and income growth in the American economy slowed down significantly, and the rising technological and economic competencies of other nations began to bother Americans, even though it is unlikely that the rising strength of other nations, particularly Japan, has been an important factor in the weak performance of the United States. To understand the reasons for the slowdown, the National Science Foundation sponsored several surveys to compare nations' innovativeness, including the Gellman survey and the Narin and Frame studies.

Gellman Study (1953–1973)

Gellman Associates tried to compare the relative innovativeness of the United States, Great Britain, West Germany, France, Japan, and Canada. An international panel of experts identified cases of innovation and ranked each of 1,310 innovations that had been introduced into the marketplace between 1953 and 1973 in terms of importance. They studied the top 500 innovations. Sixty-three percent of these came from the United States: 17 percent from the United Kingdom; 7 percent from West Germany; 7 percent from Japan; 4 percent from France; and 2 percent from Canada.

Short histories were obtained for 380 of the 500 innovations. In order to measure the innovation on a scale of relative values, the authors selected three principal categories of innovation—*radical breakthroughs, major technological shifts,* and *improvements.* Of the 100 radical breakthroughs, the United States produced 65; the United Kingdom 25; France 4; and West Germany 3. Japan produced only 2. Of the 121 improvements, the United States produced 98; West Germany 8; France 2, and the United Kingdom 2. Japan produced as many as 10 in this category.

Almost thirty years have passed since this survey. Therefore, we should ask what has happened since 1973. Furthermore, although the authors tried to make the study as objective as possible, the selection and evaluation of innovations were subjective. A more objective methodology is needed to investigate the truth of the widespread perception that Japan is an adapter rather than an innovator.

Narin and Frame Analysis (1975–1985)

With this in mind, Narin and Frame analyzed the U.S. patents issued between 1975 and 1985.[7] They found that the large number of patents issued to Japanese inventors demonstrated a burgeoning Japanese inventive vitality. From 1975 to 1985, the share of U.S. patents held by Japanese inventors doubled, from 8.9 percent to 17.9 percent. The top three U.S. patent recipients were Japanese manufacturers, and Japanese inventors obtained more patents in the United States than inventors in the United Kingdom, France, and West Germany combined.

Because the U.S. Patent and Trademark Office classifies its patents, a count of Japanese patents revealed that the areas of Japanese technological strength are concentrated in high technologies: the Japanese held 57.6 percent of the patents in photography and 43.6 percent of photocopying patents. In other technologically crucial areas the results were the same: the Japanese held 40 percent of the patents in dynamic magnetic information storage and retrieval and 43.2 percent of the patents in dynamic information storage and retrieval.

To measure the quality of Japanese innovation activities, Narin and Frame studied the examiner citations of 1975 through 1982 U.S. patents in patents granted between 1975 and 1984. Their study revealed that there was a higher proportion of Japanese-held patents in the listings of highly cited patents than we would expect statistically: 13.9 percent of all 1982 patents fell into the top 10 percent highly cited category, and 19.1 percent of Japanese-held patents fell into this category. The ratio of these two values is 1.37, indicating 37 percent more Japanese patents in the highly cited list than statistically expected. The corresponding figures are 1.06 for American-invented patents, 0.94 for U.K.-invented patents, 0.84 for French, and 0.79 for West Germany.

Table 5-1 makes visible the astonishing differences in Japan's rank-

TABLE 5-1
Comparison of the Gellman and Narin and Frame studies

	Gellman	Narin and Frame
	Radical breakthrough 1953–1973	Highly cited patent ratio 1975–1985
United States	65	1.06
United Kingdom	25	0.94
France	4	0.84
West Germany	3	0.79
Japan	2	1.37

Sources: S. Feiman and W. Fuentevilla, *Indicators of International Trends in Technological Innovation* (Washington, D.C.: National Science Foundation, 1976); and F. Narin and J. Frame, "The Growth of Japanese Science and Technology," *Science* 245 (1989): 600–605.

ing between these two studies. Using this table, we can begin to uncover the sources of these differences.

Nelson's Work

Most economic literature on convergence treats the American postwar lead as the natural result of the devastation done to its major industrial rivals by World War II and, therefore, as something that would dissolve naturally as they recovered. Nelson argues, however, that the U.S. lead had little to do with wartime devastation and there was nothing automatic about the convergence process that occurred after these rivals recovered. According to Nelson, other advanced industrial nations matched the large investments of the United States in science and engineering education and in research and development. The rise of Japan as a technological and economic power reflects this process of convergence. Japan did indeed make massive investments in education and in research and development.[8]

Nelson's argument, however, does not differentiate between Japanese innovativeness and the innovativeness of sophisticated European countries, particularly Germany. A closer look at Table 5-1 makes it clear that German ranking in innovativeness has not changed much, while the Japanese ranking has moved from the bottom to the top. Nelson's convergence theory, as good as it is, fails to explain the pattern of Japanese innovativeness. It seems quite clear that Japan became a technologically sophisticated power by taking a unique approach to innovation that is different from the ones taken by the United States and Germany.

In 1975, the Japanese created a new word, *mechatronics*, by combining the words *mechanics* and *electronics*. Essentially, mechatronics is the marriage of electronic technology to mechanical technology. From this union came a more sophisticated range of technological products, such as numerically controlled (NC) machine tools and industrial robots, as well as a series of products in which a part, or the whole, of a standard mechanical product was superseded by electronics, such as digital clocks and electronic calculators.

The diffusion rate of mechatronics technology in the case of machine tools can be measured by the ratio of NC machine tools to the total production of machine tools (see Figure 5-1). A marked increase in the diffusion rate, in fact, occurred in 1975. Since then, diffusion has been quite rapid in this industry. As a result, from 1975 to 1985, the Japanese production of machine tools rose from fourth in the world after the USSR to first (see Figure 5-2). Such growth in the world

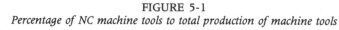

FIGURE 5-1
Percentage of NC machine tools to total production of machine tools

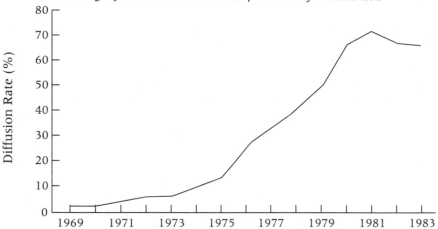

market in so short a time would not have been possible without substantial innovation in the industry, i.e., mechatronics.

In the 1980s, optoelectronics, the marriage of electronics and optics, began to yield such important commercial products as optical fiber communication systems in which the electron was united with the ephemeral photon, a particle of light, to attain greater efficiency in data

FIGURE 5-2
Production of machine tools in major economies

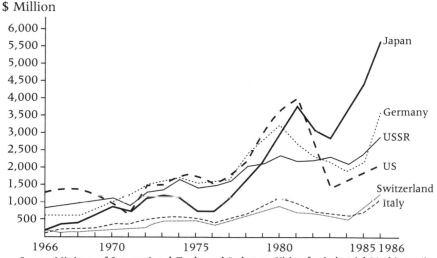

Source: Ministry of International Trade and Industry, *Vision for Industrial Machinery* (in Japanese) (Tokyo: Tsushou-Sangyou-Chousakai, 1984).

TABLE 5-2

Ranking the status of R&D among major countries

Technical Field	United States	Japan	Western Europe	USSR
Computers, chips, factory automation	9.9	7.3	4.4	1.5
Life sciences	8.9	5.7	4.9	1.3
Advanced materials	7.7	6.3	6.0	3.8
Optoelectronics	7.8	9.5	5.7	3.6

Source: G. Bylinsky, "The High-Tech Race: Who's Ahead?" *Fortune,* 13 October 1986, 18–36. © 1986 Time Inc. All rights reserved.

processing and transmission than electronics could achieve by itself. Optoelectronics has drastically revolutionized communication systems and is expected to form the next generation of information based technology.

In 1986, *Fortune* magazine asked ten scholars, business executives, government officials, and foundation leaders to rank the state of research and development in the United States, Japan, Western Europe, and the USSR on a scale of 1 to 10.[9] The study focused on four technological fields: computers, chips, and factory automation; life sciences; advanced materials; and optoelectronics (see Table 5-2). As the table shows, there is one field in which the United States was not rated number one: optoelectronics. According to the magazine, everyone agreed that the Japanese led the world in this important new technology, which was originally developed in the United States, and at least one expert noted there was little evidence that the United States would close the gap in the near future.

JAPAN'S GROWING CAPABILITY TO INNOVATE

While we have shown that mechatronics and optoelectronics are clear examples of Japanese innovativeness, Narin and Frame's count of examiner citations to nonpatent literature, most of which can be assumed to be scientific works such as books, monographs, and articles, suggests that Japanese inventions are not based as much on scientific research as those associated with Western technological efforts.[10] In order to characterize Japan's unique capability to innovate, we need to review the process of innovation in mechatronics and optoelectronics. In doing so, we will discover that the process was not a linear, or a step-by-step, technology substitution as it was when the semiconductor replaced the vacuum tube and the CD replaced the record album. Instead, the process was nonlinear, complementary, and cooperative. It blended incremental technical improvements from several previously separate fields to create products that revolutionized markets.

Mechatronics Revolution

The mechatronics revolution has transformed the machine tool industry, but the NC revolution in the Japanese machine tool industry became possible only with the cooperation of three industries outside the machine tool industry. Two inventions crucial to the machine tool industry—an improved servo motor and a compact, simple to use, and cheap to make numerical controller—are the result of Japanese innovativeness. Until these inventions were made, a huge market segment of mid-sized and small industrial customers was neglected because the NC was too costly and too large for them.[11]

The origins of machine tools with unprecedented levels of precision and reliability go back forty years. In 1952, the Massachusetts Institute of Technology invented the numerical controller. The device, however, had 2,000 mechanical valves. While it controlled a machine tool automatically, it was huge and expensive, putting it out of reach of all but the largest industries, such as aircraft manufacturers.

Having discovered a possible lucrative niche in mid-sized and small industrial customers, *Fanuc,* a spin-off from Fujitsu (a supplier of communication equipment), set out to develop a controller that was cheaper, simpler, and more compact than those of the current generation. Fanuc soon realized that a controller needs the means to execute its controls faithfully. This meant that a reliable servo motor had to be developed to position the table on which the workpiece sits as accurately as possible. In 1959, Fanuc designed a new type of servo motor, called an *electro-hydraulic stepping motor,* in which electrical control pulses are converted into mechanical movements known as *steps.* The motor removed many operational complexities and, in particular, eliminated the need for a feedback loop.

The device that harnesses the stepping motor to the worktable in a NC machine tool is called a *ball screw.* The ball screw was developed by Nippon Seiko Co. (NSK), Japan's leading maker of bearings. General Motors had been the first to use ball screws during the 1940s in the steering gear of army trucks. After General Motors' patent expired, NSK began to make ball screws. At the same time, Fanuc asked it to make a ball screw with perfect pitch. The ball screw's great advantage over its predecessor, the friction screw, lies in the lubricated ball bearings inserted between the screw's nut and bolt, which lessen friction.

Consequently, the screw's mechanical characteristics do not change during its lifetime. After all, there is little point in computers providing perfect reproducibility if the character of the parts of a machine tool can alter in the middle of a production run. Without the development of a ball screw with perfect pitch, it would not have been possible to hook up Fanuc's servo motor in an *open-loop* control system.

To complete the system, Fanuc returned to its need for a controller. The numerical controller developed by MIT in the 1950s needed a large computer capable of sophisticated calculations. Fanuc's controller, on the other hand, was based on the scientific fact that it is possible to reduce most technical drawings to arcs (which can be expressed as radii with starting and ending points) and *straight lines* (which can be defined as two points). Using this knowledge, Fanuc developed a machine that translates arcs and lines into pulses. The machine could be small because of the switch from vacuum tubes to solid-state electronics.

A further contribution came from material suppliers such as Daikin Co. A coating of *Teflon* on the sliding bed of the machine tool enabled the hook up of the servo motor, which is good for precise adjustment but weak in torque. This also made *low* speed but *uniform* movement possible, a necessity for the operation of a machine tool.

Finally, it should be noted that Fanuc's business strategy accommodated its technology development strategy. By the time Fujitsu split Fanuc off as an independent entity in 1972, the makers of machine tools had realized that numerical control gave their products a competitive edge and had begun to buy Fanuc's controllers. Because it was clear to Fanuc that if it wanted to grow, it had to provide controllers that small and medium-sized companies with fewer than 300 employees could afford, it increased production to reduce the price.

In 1986, Fanuc produced *5,000* controllers and *20,000* servo motors per month, more than half the world's production. These production levels gave Fanuc profit margins of 40 percent before tax and made it one of the most profitable companies in Japan. The number of controllers produced also made it economically feasible for the company to produce custom-made chips, which in turn enabled controllers to become even smaller. Fanuc's machines now have all their electronics on a single-printed-circuit board.

Optoelectronics Revolution

To describe recent Japanese developments in optical communication systems, we need to be familiar with Rosenberg's concept of technological disequilibrium.[12] At any time, the component parts of a machine vary in their ability to exceed their present level of performance, which is determined by the capacity of a limiting component. Thus, single improvements tend to create their own future problems, leading to further modifications and revisions. While technological disequilibrium did drive the development of optoelectronics technology, this technological revolution became possible because of the concerted effort exerted by the different industries involved.[13]

An optical communication system consists of three technological components: a light emitter (electrical to optical convertor), an optical fiber (transmission media), and a light receiver (optical to electrical convertor). For the development of optical communication systems, therefore, innovations in laser and semiconductor technologies had to be matched with innovations in optical fibers. In fact, the technological development in lasers and semiconductors progressed in an interactive and complementary manner with technological development in optical fibers, as shown in Figure 5-3.

The first optical fiber attained minimum transmission loss at the short wavelength of 0.8 μm. During the latter half of the 1970s, a

FIGURE 5-3
Technological progress in optical communications system

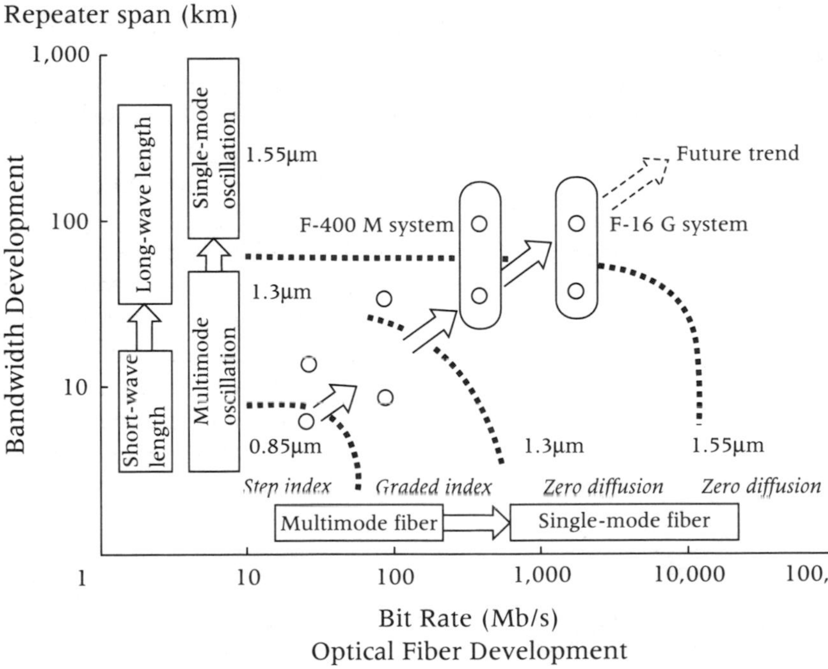

Optical Fiber Development

highly reliable semiconductor laser (GaAlAs laser), which emits light at this wavelength, was developed, and an optical communication system with SiAPD as the receiver was developed for a feasibility test of optical communication. In the meantime, a new optical fiber manufacturing technology, a more efficient vapor phase axial deposition (VAD), was developed, replacing the older, modified chemical vapor deposition (MCVD). Key to this new technology was the complete elimination of OH content, which was accomplished by developing a perfect dehydration sintering technology. As a result, the wavelength of minimum transmission loss shifted to a long wavelength of 1.3 ≈ 1.55 μm. Thus, a long wavelength laser (GaInAs laser) and receivers of Ge-APD and InGaAs-APD were developed.

All these developments contributed to the construction of the Japanese optical fiber communication line, which runs the entire length of Japan, including all the four islands. Thereafter, optical fiber of ultra small transmission loss was developed for an undersea cable line to be used for international communication. For the emitter, a *single-mode* laser (single spectrum) such as *DFB-type* and *DBRO-type* was developed. As a result, long-distance and large-capacity communication with minimum transmission loss at 1.55 μm became possible.

It is clear that technological advances were not attained simultaneously or in a parallel way in the three component areas, emitters, transmission, and receivers. Instead, innovation in one area set the technological development agendas for the other two areas. In short, "demand articulation" existed among three component technologies. Moreover, as described in chapter 3, the Japanese development of optical communications technology was conducted in a series of joint research efforts between different companies in different industrial sectors. Companies actively involved in joint efforts included NSG (glass producer), NEC (electronic device producer), SEI (cable manufacturer), and NTT (telephone company). Three of these companies—NSG, NEC, and SEI—belong to the same industrial group, the *Sumitomo* group, which may have provided a useful forum for integrating individual technology development efforts. NTT is, of course, a most important customer for the three companies.

CHARACTERIZATION: TECHNOLOGY FUSION

In every industry, the pace of technological innovation is quickening. No longer can companies miss a generation of technology and remain competitive. Adding to the pressure to keep up is the realization that innovations now cross industry boundaries. Although some companies are adept at using a variety of technologies to create new products that transform markets, many others are floundering because they rely on a technology strategy that does not work in today's environment. The difference between success and failure does not lie in how much a company spends on research and development, but in how the company defines R&D.

There are two possible definitions. A company can invest in R&D that replaces an older generation of technology—the "breakthrough" approach—or it can focus on combining existing technologies into hybrid technologies—the "technology fusion" approach.[14] Both fiber-optic communication systems and mechatronics are the result of fusion.

Historical Perspective

In his extensive studies of the history of technologies, Rosenberg has characterized *modern* technologies by their complementarities, cumulative impacts, and interindustry relationships.[15]

Because high technologies as such are not included in Rosenberg's studies, we can argue that his three characteristics only refer to embryonic high technologies. Therefore, to characterize high technologies, we need to enhance Rosenberg's characterizations. Even if high technologies embody these same characteristics, the way in which these technologies are produced has changed.

According to Rosenberg, modern technologies are characterized by complementarities, i.e., inventions hardly ever function in isolation; they depend on the availability of complementary technologies. Productivity growth is the complex outcome of large numbers of interlocking, mutually reinforcing technologies. The relevant unit of observation is seldom a single innovation, but rather an interrelated clustering of innovations. Even apparently spectacular breakthroughs usually have only a gradually rising productivity curve, but the combined effects of large numbers of improvements within a technological system may be immense.

Rosenberg's characterization of modern technology, however, is derived from long-term historical studies in which these phenomena were discovered as historical coincidences, unintended occurrences and results of passive actions. In contrast, mechatronics and optoelectronics were developed fairly quickly in a series of active joint research projects among different companies in different industrial sectors. This indicates that the institutional framework for innovation has changed.

Rosenberg argues that modern technologies are characterized by their cumulative impact. A large portion of growth in productivity is the result of the slow accretion of individually small improvements in innovations. Thus, much of the technological change that goes on in an advanced industrial economy is, if not invisible, at least of a low-visibility sort. Again, however, this characterization is derived from retrospective studies. As our case studies show, even in the early stage of the development of mechatronics and optoelectronics, the cumulative nature of innovations was recognized, and appropriate actions were taken to accommodate this effect. This indicates that the level of understanding of innovation dynamics has been enhanced.

Rosenberg's characterization of interindustry relationships notes that the benefits of increased productivity flowing from an innovation are captured in industries other than the one in which the innovation was made. A full accounting of the benefits of innovation, therefore, must include an examination of interindustry relationships because industrial development under a dynamic technology leads to wholly new patterns of specialization by both firm and industry. Thus, it is impossible to compartmentalize the consequences of technological innovation, even within conventional industrial boundaries. Based on his study of technical change in the machine tool industry from 1840 to 1910, Rosenberg suggests that the machine tool industry may be regarded as a center for the acquisition and diffusion of new skills and techniques. As a result of technological convergence, diffusion occurred in a sequential manner from firearms, sewing machines, bicycles, through automobiles.[16] The diffusion of mechatronics technologies, however, was surprisingly rapid throughout the Japanese manufacturing sector (see Figure 5-4). As shown in the figure, diffusion in several categories of machines reached saturation in less than fifteen years. Furthermore, diffusion was almost simultaneous among different manufacturing industries. This suggests that the diffusion framework has also changed.

Because the framework for innovation has changed as a result of the emergence of such high technologies as mechatronics and optoelectronics, we need to develop a new theory that encompasses Rosenberg's insights and incorporates the characteristics of high technologies. We can develop this theory through a study of the mechatronics and optoelectronics revolutions. The mechatronics revolution was generated by the fusion of mechanical technology with electronic and material technologies, and the optoelectronics revolution was generated by the fusion of glass technology with cable and electronic device technologies. Fusion is more than the combination of different technologies.

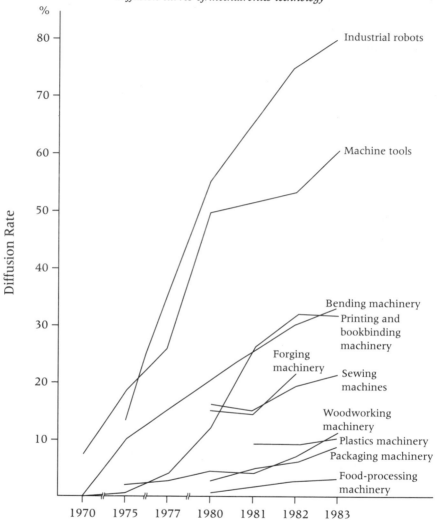

FIGURE 5-4
Diffusion curves of mechatronics technology

Source: Ministry of International Trade and Industry, *Vision for Industrial Machinery* (in Japanese) (Tokyo: Tsushou-Sangyou-Chousakai, 1984), 220.

Note: A mechatronized machine is defined as a machine with a computer control. The diffusion rate can be measured by the ratio of the mechatronized machines to the total production of the machines.

In fusion, one technology is added to another to come up with a solution greater than the sum of its parts. In other words, fusion implements an arithmetic in which one plus one makes three.

Fusion is more than complementarities because it creates new markets and new growth opportunities for each participant in the innovation. Fusion goes beyond the cumulation of small improvements because it blends incremental improvements from several (often separate) fields to create a product with some extra ingredient not found elsewhere in the market. It goes beyond interindustry relationships because, through joint research, different innovations in different industries progress in parallel with each other.

Policy Perspective

To understand the policy implications of technology fusion we must contrast technology fusion with technical breakthrough. For this purpose, we will use the transistor revolution to exemplify technical breakthrough and the mechatronics and optoelectronics revolutions to exemplify technology fusion.

First, breakthrough innovations are associated with strong leadership in a particular industry. Sometimes, they are associated with a specific individual. Fusion innovations, on the other hand, become possible through a concerted effort by several different industries. The mechatronics revolution in the machine tool industry became possible only through the cooperation of several companies, Fanuc, NSK, and Daikin. The optoelectronics revolution in optical communication systems involved such industries as glass, electronic devices, cable manufacturers, and the telephone company.

Of course, any discussion of Japan's cooperative research must include the country's industrial groups, or *keiretsu*. Some observers argue that Japan's keiretsu provide a safe environment for cross-industry R&D. The country's fiber-optics industry, for instance, owes a large part of its existence to collaborative R&D efforts within the Sumitomo group. However, the role of the keiretsu is, at most, secondary to a successful technology fusion strategy. Sharp developed its products with liquid crystal displays without the help of a powerful industrial group. Even in the fiber-optics market, the drive to create viable products was fueled by the market in the form of intense competition from companies in the United States and Europe—AT&T, Corning, Northern Telecom, Alcatel, Ericsson, Philips, and Siemens—not because of the existence of the Sumitomo group.

Second, innovation through fusion contributes to the gradual growth of all companies in the relevant industries, not to the radical growth of certain companies. The semiconductor revolution of the late 1950s, which occurred in the United States, brought about a major reshuffle in relevant industries. Many vacuum-tube manufacturers went out of business, while some new entrants, such as Texas Instruments, Motorola, and Fairchild, grew rapidly. In 1966, the market share of the new entrants was 65 percent, while that of the previous vacuum-tube manufacturers was only 26 percent (see Table 5-3).[17] This is vivid evidence of Schumpeter's creative gale of destruction. The diffusion curves of mechatronics in various categories of machines were shown in Figure 5-4. Careful analysis of this figure reveals that those categories of machinery in which the diffusion rate of mechatronics is above 30

TABLE 5-3
Semiconductor market share of major U.S. firms, selected years, 1957–1966

Type and Name of Firm	Percentage of Market			
	1957	1960	1963	1966
Western Electric	5	5	7	9
Receiving-tube firms				
General Electric	9	8	8	8
RCA	6	7	5	7
Raytheon	5	4	b	b
Sylvania	4	3	b	b
Philco-Ford	3	6	4	3
Westinghouse	2	6	4	5
Others	2	1	4	3
SUBTOTAL	*31*	*35*	*25*	*26*
New firms				
Texas Instruments	20	20	18	17
Transitron	12	9	3	3
Hughes	11	5	b	b
Motorola	b	5	10	12
Fairchild	b	5	9	13
Thomson Ramo Wooldridge	b	b	4	b
General Instrument	b	b	b	4
Delco Radio	b	b	b	4
Others	21	16	24	12
SUBTOTAL	*64*	*60*	*68*	*65*
TOTAL	100	100	100	100

Source: J. Tilton, *International Diffusion of Technology* (Washington, D.C.: The Brookings Institution, 1971), 66.

Note: b = Not one of the top semiconductor firms for this year. Its market share is included under "others."

percent are industrial robots, machine tools, bending machines, and printing and bookbinding machinery; those categories of machines that are not yet widely mechatronized are woodworking machinery, plastics-processing machinery, packaging machines, and food-processing machines.

Moreover, as shown in Table 5-4, there is a significant difference in the production growth rate for these two categories. The group with a diffusion rate higher than 30 percent has the higher growth rate. Thus, there seems to be a positive correlation between the diffusion rate and the growth rate. This indicates the possibility that the group with stagnant growth can regain momentum with the introduction of mechatronics technology.

Third, part of the breakthrough tradition in the United States stems from a defense-driven technology policy, as exemplified by the U.S. development of microelectronics and civilian airplanes. The Department of Defense funds university research that it then transfers to a limited number of defense contractors for exploitation. We might say, therefore, that breakthrough innovation is associated with defense policy.

Several authors have doubted the importance of relationships with the defense sector in the development of transistors, integrated circuits, and even aircraft.[18] Certainly, during the late 1950s and early 1960s, a considerable portion of the funding in these areas came from the Department of Defense, while in other areas funding came from NASA. At the same time, however, people interested in civilian markets were

TABLE 5-4

Mechatronics ratio and growth rate

Category of Machinery	Percentage of Mechatronized Machine	Growth Index of Production 1983/1977
Industrial robots	80	10.67
Machine tools	60	2.52
Bending	30	7.00
Printing & bookbinding	30	2.44
Forging	20	0.76
Sewing machines	20	0.99
Woodworking	10	1.35
Plastics	10	2.18
Food processing	3	1.71

Source: Ministry of International Trade and Industry, *Vision for Industrial Machinery* (in Japanese) (Tokyo: Tsuschou-Sangyou-Chousakai, 1984), 220.

investing in these areas.[19] At the very least, therefore, we can say that this relationship with the defense sector is subtle and indirect.

Industrial policy can also induce innovation through fusion. In Japan prior to 1971, the Temporary Measures for the Development of Machinery Industry Law (1956) and the Temporary Measures for the Development of Electronics Industry Law (1957) were enforced separately. Each law provided low-interest loans to companies for technology development projects, particularly projects with commercial possibilities, from semigovernmental financial institutions, such as the *Japan Development Bank*.

In 1971, these two laws were transformed into the single Temporary Measures for the Development of Specified Machinery and Electronics Industries Law (see Figure 5-5). The minister of MITI explained the objectives of the laws as follows:

> In deciding upon an Intensification Plan, the competent minister shall take note of the increased interrelationships among different industries,

FIGURE 5-5
Chronology of mechatronics-related laws

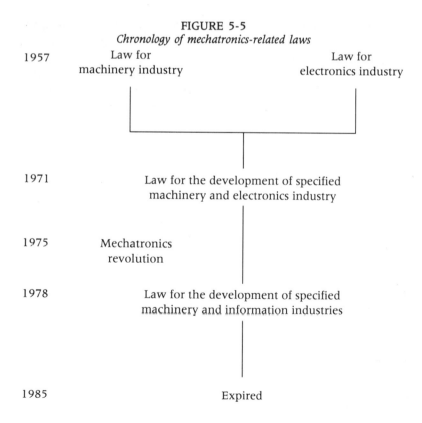

1957	Law for machinery industry	Law for electronics industry
1971	Law for the development of specified machinery and electronics industry	
1975	Mechatronics revolution	
1978	Law for the development of specified machinery and information industries	
1985	Expired	

in particular, between the machinery and the electronic machinery and equipment, and shall pay due consideration to the direction of so-called *consolidation* of machinery and electronics into one or *systematization* of them.[20]

Thus, the mechatronics revolution was clearly envisioned in this law. In 1978, this law was transformed into the Law on Extraordinary Measures for the Development of Specified Machinery and Information Industries, which expired in 1985. Although there are doubts about how much these laws influenced industry behavior, the merger of the two laws clearly showed where the government believed major innovations were to occur in the 1970s.

Government involvement in the optoelectronics revolution can be seen in the *HI-OVIS* system project, a community communication system using optical fiber, which was sponsored by MITI and promoted by Sumitomo Electric, Fujitsu, and Matsushita. The project was completed in 1978. We might say, therefore, that the initial development of the optoelectronic revolution was created by government-sponsored projects. At about the same time, however, the NEC-SEI group from Japan completed the Disney World project in the United States, which then purchased the optical fiber on a commercial basis for the first time. Therefore, we can say that the initial market was the entertainment business, obviously not the most serious customer in terms of cost, reliability, scale, and so forth.

While it is true that the government (MITI, in particular) actively promoted technology fusion through legislation and govenment-funded research projects, its role is secondary in successful technology fusion. In the evolution of mechatronics and of optoelectronics, neither the laws nor MITI's demonstration project materially changed industrial behavior. Research and development would have taken place with or without the government's prompting.

A GRAPHICAL REPRESENTATION

Fusion grows out of long-term R&D ties with a variety of companies across many different industries. Thus, investment in joint R&D and partnerships must go beyond tokenism. Although the process of technology fusion begins with an industry's interests in product fields outside its principal fields, the traditional dichotomy between product and process innovation is not helpful in analyzing technology fusion. A more relevant dichotomy is a sector's industrial R&D within and outside its principal product fields because industrial R&D activity outside principal product fields is directed toward creating technology fusion.

Once again, the Statistics Bureau's *Survey of Research and Development,* which disaggregates companies' intramural expenditure on R&D into thirty-two different product fields, makes it possible to turn the concept of technology fusion into an empirical study.[21] Of course, not all industrial R&D activity outside principal product fields involves technology fusion. To qualify as fusion, cross-industry R&D must be both *reciprocal* and *substantial.*

Reciprocity

Reciprocity is the essence of technology fusion. All participants in a joint research project must enter as equals (mutual respect); each must assume responsibility for contributing a certain expertise (mutual responsibility); and all must share in the success of the development (mutual benefit). The ceramics industry provides a good example of the reciprocity of technology fusion. Although companies like Kyocera invested heavily in research for new packaging materials for electrical and electronic equipment, it was not until electrical and electronic industries invested jointly with ceramics companies that a new generation of industrial ceramics emerged as a technical field, to the benefit of all involved.

Fusion also occurs when a new company or group of companies from a new industry enters the scene. The fiber-optic cable developed by Nippon Sheet Glass in the 1970s lacked mechanical strength, and the quality of transmission over long distances was poor. SEI developed a coating technology that strengthened the cable, which solved the mechanical fragility problem. A communications service company and a cable manufacturer then solved the transmission-loss problem through a joint research effort using longer wavelengths.

Thus, the minimum unit of technology fusion is a pair of two different industrial sectors, *i-th* industry and *j-th* industry, for example. The unit of analysis is a sector's R&D investment into another industry's principal product fields, for example, *i-th* industry's R&D investment into *j-th* industry's principal product fields. This, however, only describes one-way investment. It may be only a diversification effort and may not lead to innovation.[22]

If the investment is done *reciprocally,* it will lead to the creation of a new technological area and possibly to innovation. Although reciprocal investment may not be sufficient for technology fusion, it is obviously necessary. On this basis, we can assume that technology fusion between *i-th* industry and *j-th* industry is realized only after the investment is done in two ways, i.e., *i-th* industry's investment into *j-th* industry's principal product fields and *j-th* industry's investment into *i-th* industry's principal product fields.

It could be argued that if reciprocal investment is simply the mixing together of different types of technology, it is possible to achieve technology fusion by merging corporations that possess different types of technology. Such a combination through mergers, however, merely involves one technology using another with little likelihood of the technology fusion discussed here. If reciprocal investment takes the form of two corporations undertaking joint research, the two technologies are of equal importance and technology fusion can easily result. The hypothesis of reciprocity is intrinsic to technology fusion.

Substantiality

Substantiality means that management makes a commitment—from early exploratory research work to advanced product development—to the joint R&D project. Substantiality gives the company's partners and employees assurances that the funds will be there to see the project through to completion once senior management has committed to it.

Considering the dynamics of R&D programs described in chapter 3, we can assume that research reaches the *development* state only if the expense exceeds a certain amount. What is important for the purpose of this chapter is the demarcation between the *exploratory state* and the *development state*. In Chapter 3, the state transition model of an R&D program's investment is based on the concept of freezing rate function $r(C)$, i.e., a probability that an R&D program's annual expenditure is frozen at C, given that it has come to the investment level of C. By assuming that the freezing probability function $r(C)$ is a decreasing function of C, I can make a demarcation as follows: an R&D program in the development state can be defined as one whose freezing probability is smaller than a specified value r^*. Based on each industry's estimated relation between the freezing probability (r) and the annual expenditure (C), as identified in chapter 3, we can identify the boundary value of annual expenditure beyond which an R&D program can be supposed to enter the development state, as shown schematically in Figure 5-6. The boundary value for industrial chemicals cannot be obtained because we have classified this as a science-based industry. In calculating the boundary value for it, therefore, we used the freezing

FIGURE 5-6
Calculation of threshold value for i-th industry

Freezing rate

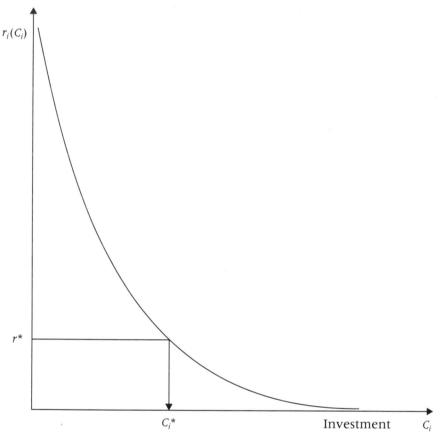

rate function for the dominant-design pattern, even though the level of significance for the curve-fitting is not satisfactory. Thus, for each industrial sector, we can obtain the threshold value C_i^*, which demarcates exploratory search from development.

Fusion Graph

Based on the two requirements of reciprocity and substantiality, we can now formulate a model of the realization mechanism of technology fusion.

Given R_{ij} (*i-th* industry's annual R&D expenditure in the *j-th* industry's principal product fields), then, technology fusion is assumed to be *realized* between the *i-th* and *j-th* industry, only if

$$R_{ij} > C_i^*,$$

and

$$R_{ji} > C_j^*.$$

Thus, we can construct a model of technology fusion that is associated with a given pair of industries and move on to analyzing the network structure of technology fusions. The best way to describe the structure of technology fusion is to represent it graphically. In a *nondirected graph*, each industry is represented by a *vertex*, and an *arc* is drawn between a pair of vertices only if there is realized technology fusion between the two industries represented by the vertices. We can call this graph a technology fusion graph.

We can formulate a technology fusion graph by a nondirected mathematical graph, $F = [F_{ijt}]$, where

$$F_{ijt} = \begin{bmatrix} 1 & \text{if } R_{ijt} > C_i^* \text{ and } R_{jit} > C_j^*, \\ 0 & \text{otherwise.} \end{bmatrix}$$

As the Statistics Bureau's survey covers every year since 1970, a graph can be constructed for each year since 1970. To do so, however, we have to determine the critical value of the freezing rate (r^*), which demarcates the exploratory and development states. Because there is no information available, it can be determined only on an empirical basis. In other words, it can be calibrated in the critical freezing rate. We know that the Japanese created the word "mechatronics" in 1975, and by measuring the ratio of the NC machine tools to the total production of machine tools (Figure 5-1), we can see, as far as machine tools are concerned, a marked increase in the diffusion rate of mechatronics technology occurred in 1975. However, mechatronics encompasses not only NC machine tools but also industrial robots and digital clocks. Therefore, we can identify the realization of the mechatronics revolution by the persistent existence of the *quadruple* connection among ordinary machinery, precision instruments, electrical machinery, and communications and electronic equipment, i.e., all possible arcs between every pair of the two industries out of the quadruple (Figure

5-7). Now, we can determine the critical freezing rate in which the quadruple connection begins to appear only in 1975 and is maintained after that. Before 1975 we will have only imperfect quadruples, but after 1975 we have complete quadruples. Through this kind of calibration, we find the critical value of the freezing rate $r*$ should be set at 0.07. We can then assume that the R&D program with a freezing rate smaller than 0.07 is in the development state, and on this basis, we can calculate the threshold value of annual R&D expenditure for each industrial sector. The results of this calculation are shown in Table 5-5. Once the threshold values for all industries are determined, we can draw a technology fusion graph for each year since 1970. In this graph, as in the fusion graph for mechatronics, technology fusions are clustered around three major industries—ordinary machinery, electrical machinery, and industrial chemicals. The first two industries belong to the fabrication industry, while the last belongs to the material industry. Therefore, we can design the fusion graph so that the vertices representing industries clustered around ordinary machinery are located at the top corner of the graph, those that are clustered around electrical machinery at the right-hand corner, and those that clustered around industrial chemicals at the bottom-left corner.

In the fusion graph for 1970 depicted in Figure 5-8, the three clusters are separated from each other. This implies that there was no technology fusion among these three clusters at that time.

THE STRUCTURE OF HIGH-TECH FUSION

Using these technology fusion graphs, we can show that most high technologies are the products of fusion innovation. By reviewing the time change in the network structure depicted in a fusion graph, we can see how such technologies as biotechnology, mechatronics, and new ceramics were developed as high-tech areas.

First, we need to review the formation process of mechatronics, depicted in Figure 5-7. Technology fusion between ordinary machinery and electrical machinery began in 1971. Around 1973, communications and electronics joined this grouping, forming the triple connection among the three industries, which indicates an early version of mechatronics. In order to perfect mechatronics, however, precision instruments had to be included. Thus, it was not until 1975 that the quadruple connection began to appear persistently among ordinary machinery, precision instruments, electrical machinery, and communications and electronic equipment.

FIGURE 5-7
Technology fusion graph for mechatronics, 1972

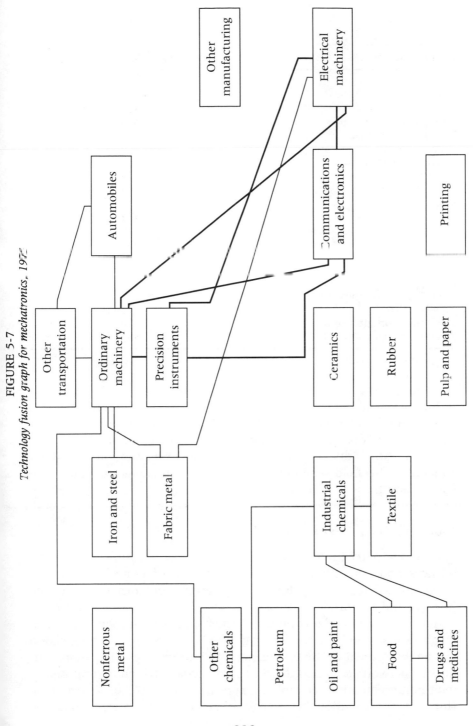

TABLE 5-5
Identification of threshold values

Industrial sector	Threshold value (¥ million)
Food	467
Textile	822
Pulp and paper	365
Printing and publishing	626
Industrial chemicals	1834
Oil and paints	579
Drugs and medicines	300
Other chemical	648
Petroleum and coal	369
Rubber products	695
Ceramics	984
Iron and steel	639
Nonferrous metals	1003
Fabricated metal	773
Ordinary machinery	656
Electrical machinery	231
Communications and electronics	355
Motor vehicles	948
Other transportation equipment	650
Precision equipment	382

In fact, Fanuc's president, Seiuemon Inaba, who orchestrated this technology fusion, started his career as a precision instrument engineer. Inaba carefully investigated all the available machine tools and made rounds of the tool companies to ask about their requirements. He appeared on NSK's doorstep, urging it to make a precision ball screw for Fanuc. He also held seminars to explain the advantages of numerical control and how toolmakers could apply it.[23]

Biotechnology Revolution

The concept of technology fusion can be extended beyond the physical sciences and chemistry. With the emergence of biotechnology, the trend toward technology fusion has become obvious even outside physical and chemical sciences.

In 1974, a triangular connection emerged between food, drugs and medicines, and industrial chemicals (see Figure 5-9). This connection can be interpreted as the emergence of biotechnology. The dual connection between food and drugs and medicines, which is supposed to indicate fermentation technology, appeared in 1970 (see Figure 5-8). However, fermentation technology alone lacked the components of a

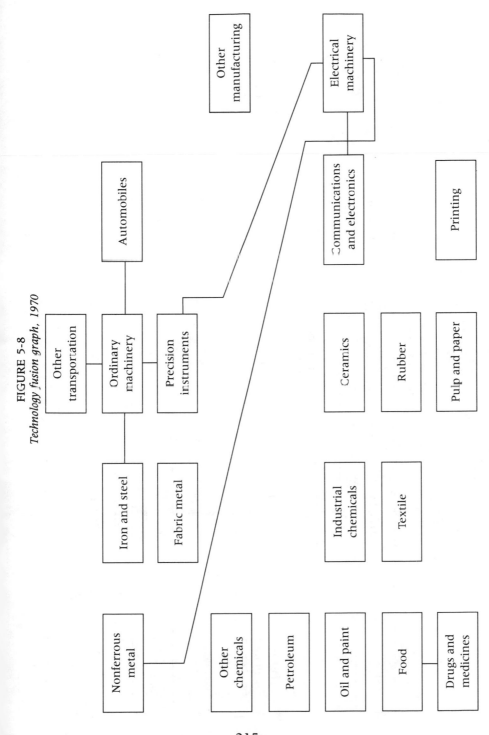

FIGURE 5-8
Technology fusion graph, 1970

215

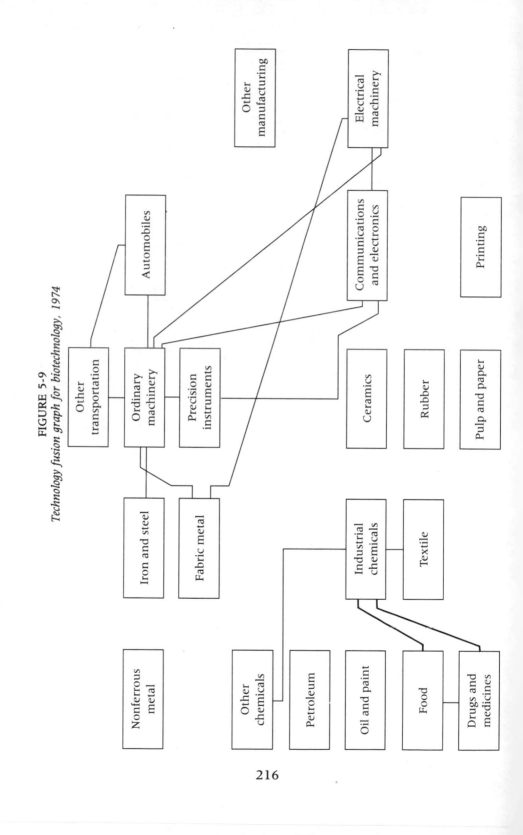

FIGURE 5-9
Technology fusion graph for biotechnology, 1974

216

science-based industry as we defined it in chapter 3. Thus, when the industrial chemicals industry was added to the network, biotechnology was established as a high-technology area.

The emergence of biotechnology has led to the "rational drug design" phenomenon, which promises to streamline and enhance drug development and reshape the biotechnology and pharmaceutical industries.[24] The traditional way of developing pharmaceuticals was to screen thousands of chemicals in an inefficient, time-wasting hit-or-miss search. Drug designers today use biotechnology to help them work backward from what biologists know about a disease and how the body fights it. Dozens of companies zeroing in on drugs to fight nervous system disorders have used biotechnology to unravel brain function, and they will probably use chemical synthesis to create their drugs.

Although the development of biotechnology is targeted toward producing pharmaceuticals in the United States, the Japanese think it has a broader area of application. Biochemistry is taken for granted in Japan because of its long tradition in fermentation technology, which was discussed in chapter 2, where it was noted that the creation of amino acid industry was accomplished by following a trickle-up business and technology diversification. Japan's familiarity with this component of biotechnology may explain why it is said that Japan will become a front-runner in a much shorter time than one might expect.[25]

New Materials Revolution

A review of Figures 5-7 and 5-9 will show that during the 1970s, technology fusions occurred only within the fabrication industries (mechatronics) and within material-related industries (biotechnology). There was, at this point, no fusion between the materials and the fabrication industries.

Since 1980, however, fusions have appeared frequently between these two clusters of industries. These fusions indicate the innovation pattern of the so-called new materials revolution. Throughout history, materials have given their names to epochs: the Stone Age, the Bronze Age, the Iron Age. Until recently, the materials involved have been nature's gifts or simple improvements on them. Now, however, we are on the threshold of the age of man-made materials. Today, scientists can tailor the basic structures and properties of materials to suit their needs.[26]

For many Japanese, these materials represent the coming of the age of "fourth-generation" materials. First-generation materials are stones and woods, which are used primarily in their raw form. Second-generation materials are copper and iron, which become usable by extracting components from naturally available materials. Third-generation materials are plastics, which are not available in nature but are synthesized artificially. For the fourth-generation, engineers will custom design new materials for specific uses by manipulating atoms and electrons.

The essence of the new material revolution is the technology fusion between two clusters of industries, fabrication and materials, which had never been realized before.

In 1982, technology fusion between ceramics and ordinary machinery and between ceramics and electrical machinery appeared (see Figure 5-10). These connections represent the emergence of new ceramics. The connection between ceramics and ordinary machinery, which represents *fine ceramics,* was formed in 1980. In 1981, the connection between ceramics and industrial chemicals, indicating the emergence of the new material, was formed. Thus, in 1982, when the connection between ceramics and electrical machinery was added, a new materials revolution was realized as far as ceramics is concerned.

The pattern of this revolution probably will not follow the conventional pattern of a materials revolution. The main actors in the materials revolution will not be the materials industry, but rather manufacturers that use materials technology to solve specific customer problems. Just as NTT "pulled" fiber-optics development to satisfy market demand, so will a new breed of manufacturing companies pull materials technology to the market in order to differentiate their products. Aeronautical engineers and semiconductor engineers, for example, will design materials for fuselages and computer chips, based on their customers' needs for flexibility, strength, conductivity, environmental stress, and a host of other critical factors.

If this happens, Japan's materials industries will no longer be considered weaker than its fabrication industries. The results of the *Fortune* scoreboard (see Table 5-2) provided an early indication of the turnabout. According to these results, the ratio of Japanese/U.S. scores was as close as 0.82 in advanced materials, while the ratios in computers and life sciences were 0.74 and 0.64.

Many leading technology fusion companies in Japan are already taking steps to harness the power of this new generation of materials. As Tadahiro Sekimoto, the president of NEC, said, "The company that controls materials development will dominate in the electronics industry."[27] Indeed, experts in various fields from several countries now acknowledge the importance of the materials industries. Professor Hiroshi Inose, winner of the Marconi Award for his work in digital

FIGURE 5-10
Technology fusion graph for new ceramics, 198☰

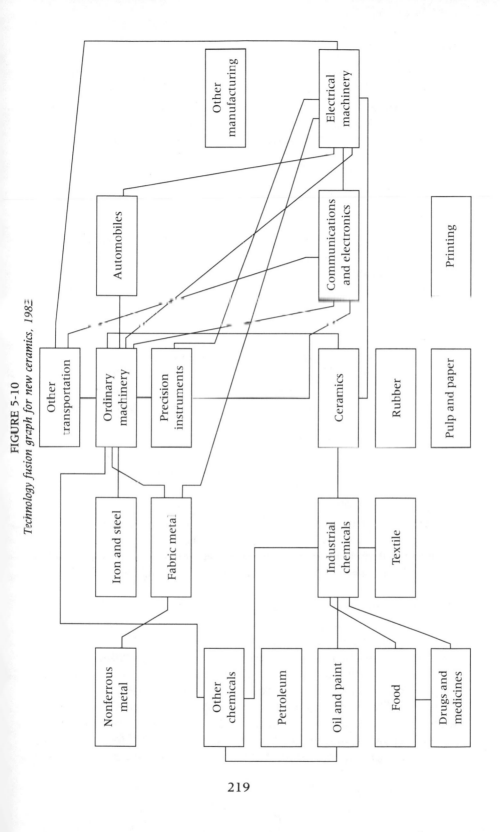

communications, claims that the focus of technological development, which once shifted from systems to devices, has shifted again to materials. In the United States, Ralph E. Gomory, former senior vice president and chief scientist at IBM, has pointed out that every single step in computing has depended upon solving one materials problem after another.[28]

Fusion beyond Manufacturing

As more and more companies accept and make fusion a part of their overall technology strategies, it will play an increasingly important role in product development. This will open the floodgates to even more cross-industry R&D. In the 1970s and 1980s, technology fusion was limited to manufacturing industries. In the future, fusion will go beyond manufacturing.

One twenty-year product development vision in the Japanese entertainment industry involves what Japanese engineers call "media design," a fusion of audio and video hardware and software with the entertainment industry's creativity and artistry. One product concept now being toyed with is interactive cinema in which a moviegoer dons a lightweight headset, fitted with a pair of goggles, and slips on a pair of electronic gloves, thus becoming an actor in a virtual-reality movie. By shrinking the components and bundling them into an affordable home unit, the virtual-reality theater could become the twenty-first century's equivalent of the videocassette recorder.

Although this idea sounds like pure science fiction, companies like Sony and Matsushita are taking it seriously. They are setting long-term research and business agendas that they expect will deliver a new generation of such innovative products within the next two decades. Both companies have already taken the first steps toward fulfilling this vision: Sony acquired Columbia Pictures Entertainment in 1989, and Matsushita purchased MCA Inc. in 1990. As Michael Schulhof, vice chairman of Sony Corporation of America, noted, "The acquisition of a major film studio extends Sony's long-term strategy of building a total entertainment business around the synergy of audio and video hardware and software." Akio Morita, chairman of Sony, confirmed Sony's strategy when he asserted that the possibilities and synergies created by the merger of Japanese hardware and American software were already yielding new products.[29] Both purchases open opportunities for replacing analog production systems using film with digital systems using computer-generated images. Once the industry is all digital, a whole new set of computer-based innovations, such as high-definition television and digital audiotape technology, are feasible.

TECHNOLOGICAL FORECASTING

Our method of fusion graph analysis could become an empirical technological forecasting method. It would be, however, totally different from conventional methods, such as the *Delphi* method, which rely heavily on the subjective human judgment of experts.

If high technology is a product of innovation that comes from fusion, reciprocal investment patterns, which act as signs of future technological fusion, could be used as leading indicators of high-technology innovation. Even if the forecasting method is based on the subjective judgment of experts, we can design the forecasting panel so that the reciprocity elements of high technology are built into the forecasting process. In order to demonstrate this possibility, we will review two surveys made in Japan.

Expected Technological Advances

Roughly every five years, MITI draws up a list of major fields of technology. At the last such meeting, scientists and engineers representing seven technological fields (new materials; biotechnology; electronics; information processing; energy; aerospace; and construction and transportation) gathered together to draft a latest list. This time, however, a somewhat different approach was tried.

Seventy scientists and engineers working in corporate R&D and planning in each field were asked to list the fields outside their own from which they most eagerly anticipated new advances.[30] They were also asked about their expectations in the short term (0–5 years) and in the long term (5–10 years). In all, the questionnaire was sent to 490 people in 231 companies. As shown in Table 5-6, a total of 149 responses was obtained, a return ratio of 30 percent. The most responses came from information processing, which had a return ratio of 42 percent. Energy was second with 33 percent. These ratios reflect the intrinsic nature of these particular technical fields, which are heavily dependent upon advances in other technical fields for their own advances.

Table 5-7 summarizes the results of the survey for the short term (0–5 years). The next to last row gives each field's maximum possible number of responses, i.e., the total number of respondents minus the

TABLE 5-6
Number of respondents by field

Technological Areas	Research/ Development	Corporate Planning	Total	Response Ratio
New materials	15	6	21	30.0%
Biotechnology	7	12	19	27.1
Electronics	13	6	19	27.1
Information processing	21	9	30	42.1
Energy	18	5	23	32.9
Aerospace	8	10	18	25.7
Construction/Transportation	11	8	19	27.1
TOTAL	93	56	149	30.4%

number of respondents in the field itself. Expectations are reported in two rows: Total (A) records the number of respondents in other fields who indicated high expectations for developments in that field; (A/B) expresses this number as a percentage of the maximum possible number of respondents.

The survey's results show that within the next five years the greatest expectations are for electronics. This is followed by new materials and then by information processing. Relatively little is anticipated from the fields of biotechnology, aerospace, energy, and construction.

The high marks for electronics do not need explanation. The high score for information processing reflects the fact that it is expected to provide important tools for research and analysis. Expectations vis-à-vis biotechnology remain low, however, as this area has yet to move much beyond the conceptual stage.

The low expectations for aerospace and energy, however, merit particular attention. It is widely known that state-sponsored aerospace and energy projects, many of which are defense-related, have produced numerous breakthroughs in the form of spinoffs in new materials, electronics, and software. Therefore, the low expectations for this field seem to indicate that the development of new materials and computers is expected to occur in direct response to specific technological needs rather than emerging from defense-related projects.

In order to determine the shift from the short term (0–5 years) to the long term (5–10 years), expectation scores for the two periods are compared in Figure 5-11. As seen in the figure, great expectations are held for new materials in the long term, i.e., 94 positive responses or 73 percent of all possible responses. Expectations for electronics,

TABLE 5-7
Expectations for development across technological fields

Expectations by	Expectations for						
	New materials	Biotechnology	Electronics	Information	Energy	Aerospace	Construction/transportation
New materials	—	2	15	10	11	9	2
Biotechnology	12	—	12	14	0	1	1
Electronics	16	1	—	13	4	6	2
Inf. processing	11	3	29	—	2	2	1
Energy	19	2	18	14	—	3	6
Aerospace	13	3	11	9	4	—	2
Construction/ transportation	14	3	9	12	6	2	—
TOTAL (A)	85	14	94	72	27	23	14
No. of possible respondents (B)	128	130	130	119	126	131	130
(A/B)	66%	11%	72%	61%	21%	16%	11%

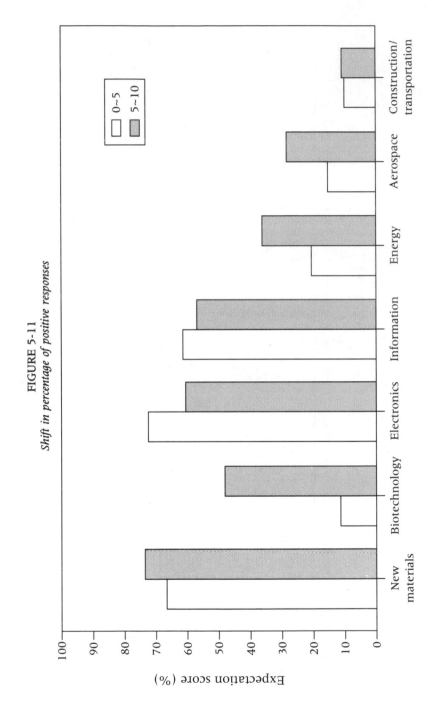

FIGURE 5-11

Shift in percentage of positive responses

however, fall to second place. This reflects the widely held view that ultimate problem-solving will be effected only by the development of new materials, not by application of electronics. The increasing expectations for materials technology reflect, at least in part, a dramatic change in the nature of technological innovation in the materials field.

Increased expectations for biotechnology in the long term reflect the hope that biological processes will replace physical and chemical processes in various fields of manufacturing. The score for biotechnology increases from 14 points (11 percent) in the short term to 62 points (48 percent) in the long term.

New Trends in Research Consortia

The shift to technology fusion is reflected in the organization of Japanese research consortia today. An analysis of membership reveals that the average number of industries per project is increasing, while the number of participating companies per industry is decreasing. In other words, collective research in Japan is beginning to bring together companies from different industries rather than different companies within the same industry.

To reach this conclusion, we reviewed the thirty-year history of ERAs. After excluding those associations that companies with stock registered in the open stock market did not join, eighty-eight ERAs were selected for study.[31] For the entire set of ERAs, the average number of industrial sectors per project was 3.6, and the number of participating companies per project was 3.2. The ERAs were divided into six cohorts based on the date of establishment. The time periods for the six cohorts are shown in Table 5-8. The average number of industrial sectors represented in the thirteen ERAs established between 1970 and 1974 was 2.2 sectors per project. This number has increased steadily. There were 4.3 sectors per project in the twenty-three research associations established after 1985. On the other hand, the average number of participating companies per industrial sector has held steady at about three.

It would appear that government-organized research consortia lead to the networking of different kinds of "core competence" owned by different companies in different industrial sectors.[32] Technology policy, then, plays the key role of matching competence and needs, thus fostering the possibility of fusion. Because these networks are formed more quickly than they would be if the initiative were left entirely to the firms themselves, the technical and market information exchanged

TABLE 5-8
*Classification of member enterprises by industrial sector
per engineering research association*

Time Period	Number of Projects	Number of Industrial Sectors per Project	Number of Registered Companies per Industrial Sector
1961–1964	3	3.0	3.2
1965–1969	0	—	—
1970–1974	13	2.2	3.0
1975–1979	14	3.4	3.4
1980–1984	35	3.9	3.3
1985–	23	4.3	3.0
TOTAL	88	3.6	3.2

Source: I. Shirai and F. Kodama, "Quantitative Analysis on Structure of Collective R&D Programs by Private Corporations in Japan," NISTEP Report No. 5 (in Japanese) (Tokyo: National Institute of Science and Technology Policy, 1989).

through these relationships lead to faster corporate investment in innovation.

GLOBAL PARTNERING

Technology fusion is a substantial competitive advantage for Japanese companies today. Will this advantage result in worldwide share gains by Japanese industry, or will technology fusion foster global partnering? After all, the need for multitechnologies capability may force cooperation among high-tech companies in different industries, even when they are based in different countries. It seems unlikely that any one national industry will develop hegemony across the varied and complex technologies of the high-tech industry. Therefore, global partnerships are very possible.

Since the emerging technology fusion paradigm is characterized by reciprocity, global partnering could be very effective, because a partnership is, by definition, a *two-way* street. Yet, because of political posturing and suspicion as well as ignorance about the process of the high-tech development, international business has not reaped the benefits that could be gained from global partnering.[33] Technology fusion, however, is leading to a concerted effort to pool complementary talents in an interactive way and to the creation of genuine partnerships that have real, tangible benefits for all the participants.

The forms these new global partnerships may take are complex and still evolving. In what follows, however, we will discuss two modes of global partnering that allow companies to capitalize on the emerging technology fusion paradigm: one based on complementary relations between two large companies; the other promoted by government policy.

Complementary Assets

To introduce the concept of "complementary assets," David Teece noted that the successful commercialization of an innovation requires that the innovation is utilized in conjunction with other capabilities or assets.[34] Marketing, competitive manufacturing, and after-sales support are almost always needed. These services are often obtained from specialized complementary assets. When the innovation is systemic, the complementary assets may be other parts of a system. In industries in which technological change requires deployment of specialized and/or cospecialized assets, the firm boundaries established prior to change may no longer be efficient. The shift toward technology fusion is making existing company boundaries rather fragile. Indeed, technology fusion may result in fusing one company with another, and this could occur internationally.

Today, many American and Japanese firms see strategic alliances as alternatives or supplements to other measures for reducing cost and risk. This is not the result of a sudden change in philosophy, but rather the result of a gradual change over the postwar period that reflects changes in technology, commercial competition, and positions of U.S. and Japanese industries, especially the semiconductor industries. Many U.S. firms in the semiconductor industry have come to see strategic alliances as a means of eliminating the need to make certain manufacturing, marketing, or R&D investments that they might otherwise have to make.

To understand the scope and nature of the change, we have only to look at earlier alliances. From 1950 to 1980 very few strategic alliances were concluded. The few that existed were in the form of simple licensing agreements, involving the sale of basic U.S. patents to latecomer companies in Japan. During the early 1980s, however, the frequency and complexity of alliance formation increased markedly, especially in the area of U.S. licensing of memory and microprocessor technology. In the mid-1980s, a number of agreements were signed in the area of semiconductor equipment, and by the late 1980s, the proliferation of U.S.-Japan alliances had reached a peak.

According to D. Okimoto and his colleagues, in the mid-1980s, when the number of strategic alliances soared, the impetus grew to pool resources to fuse one company's strengths with those of a partner and thus consolidate and expand competitive advantages while offsetting company weakness.[35] The Toshiba-Motorola alliance is a typical example (see Figure 5-12).

The heart of the relationship was the exchange of Motorola's microprocessor technology for Toshiba's DRAM design and manufacturing technology. By bringing together complementary assets, the partners filled out their product portfolios, enhanced their overall technological capabilities, and compensated for in-house gaps and weaknesses. Since late 1986, the semiconductor partnership between Motorola and Toshiba has evolved in the direction of deeper interaction and a broader range of collaborative activity. The alliance is sometimes cited as an example of a U.S.-Japan linkage in which the benefits to the partners have been relatively evenly balanced.

Motorola's alliance with Toshiba should be seen in the context of the company's effort to gain access to the Japanese semiconductor market. In 1962, Motorola set up a sales office in Japan, but it soon became clear that this approach would not lead to significant participation in the Japanese market. In 1970, Motorola set up a joint venture with Alps Electric Co., an electronic component maker it had previously dealt with in other product lines. The joint venture, Alps-Motorola Semiconductor K.K. (AMSK), did back-end assembly and testing of devices, Japanese sales, and warehousing. Motorola hoped that AMSK would evolve into a major Japanese supplier of its products, but the venture ran into a number of difficulties and collapsed in the

FIGURE 5-12
Complementary relationships in the Toshiba-Motorola partnership

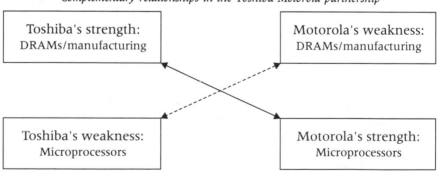

Source; D. Okimoto et al., *U.S.-Japan Strategic Alliances in the Semiconductor Industry: Technology Transfer, Competition and Public Policy* (Washington, D.C.: National Research Council, 1992), 29. Reprinted by permission.

1974–1975 industry recession. While the business climate did not improve the venture's chances, failure may have been due in part to Motorola's inexperience in operating in Japan. Motorola learned that it was probably best to leave personnel and other administrative functions in the hands of Japanese partners while retaining authority over operational matters.

Around 1980, Motorola realized that the continuing growth of the Japanese industry necessitated another effort to gain access to the market and to develop a manufacturing capability. A semiconductor company without a significant foothold in Japan would miss a large chunk of the world market. In addition, Japanese customers were the most demanding in the world, and lessons learned from marketing in Japan could be applied to Motorola's global operations. Motorola decided to try another joint venture and chose Toko, Inc., as a partner. Toko was a small electronic component maker—primarily radio parts. The interests of the two companies were complementary, and the alliance lasted about two years, after which Motorola acquired the joint venture and renamed it *Nippon Motorola*. The subsidiary, which does wafer fabrication and packaging for the Japanese market, is the world supplier of several Motorola products. Nippon Motorola made a major effort to meet the needs of the Japanese market, but progress was slow and difficult.

In the mid-1980s, Motorola realized that it needed to change tactics again to strengthen its position in Japan. By the end of 1985, Motorola stopped making DRAMs in the United States for merchant sale, as had all U.S. merchant companies except Texas Instruments and Micron Technologies. Before approaching any Japanese companies, Motorola evaluated possible partners by looking at both business and nonbusiness factors. Toshiba rose to the top. Its partnerships with foreign companies had generally gone well. Both sides usually benefited, and it was thought that Toshiba could protect proprietary technology from other Japanese companies. For Toshiba, the partnership was attractive because of Motorola's leadership in microprocessors. While Toshiba had moved past NEC to become the leading Japanese maker of DRAMs, it lagged behind NEC and Fujitsu in microprocessors. An alliance with Motorola would help redress that deficiency.

The first element of the partnership, which was announced in the press in August 1986, was an agreement for Motorola to purchase DRAM dice fabricated by Toshiba. In December of that year, the two companies announced an agreement to engage in negotiations to extend the relationship further. There were a number of advantages for Motorola. Among other things, in addition to the basic swap of technology for market access, Motorola would gain access to DRAMs manufactured at the new facility and could offer its customers a full product line. Furthermore, Motorola could access Toshiba's DRAM manufacturing know-how for use in its other fabrication facilities. The partners designated a site for the joint venture, Tohoku Semiconductor, in May 1987. The facility, located at Sendai in Tohoku area, was completed about a year later. Motorola began by transferring know-how concerning its 8-bit microprocessor. There was an understanding that the technology transfer would be extended to more advanced products as sales milestones were hit.

Microprocessor technology transferred to Toshiba through the alliance consisted of circuit design and software operating codes. Toshiba became an original equipment manufacturer (OEM) for Motorola's 68020 microprocessor (a 32-bit product) in 1988. In May 1989, the two companies agreed to expand cooperation at the Sendai facility to 4-megabit DRAMs. In June 1990, the two companies announced plans to codevelop a microprocessor for a new Toyota engine. They also agreed to collaborate on developing semiconductors relevant to high-definition television (HDTV). Both companies have complementary assets to influence the development of HDTV in the Japanese and American markets.

The Motorola-Toshiba alliance has had mixed results in some business areas, but the relationship exhibits a logic that extends beyond short-term business results: the swap of microprocessors for DRAM technology as well as marketing collaboration. Both firms are strategically sophisticated and long-term oriented. There are areas in which their linkage may develop in interesting ways, such as complementary approaches to systems. One market that will probably grow in coming years is that resulting from the fusion of cellular telecommunications and portable computing. Motorola's technical strengths in the former—including its efforts to develop a global cellular network linked by satellites—and Toshiba's technical strengths in the latter point to promising possibilities for collaboration, but to date no agreements have been announced in this area.

Government Policy

We have already noted the evolution of Japan's government-sponsored R&D consortia in recent years, which reflects a technology policy that recognizes and encourages technology fusion. In the past several years, moreover, the Japanese government has indicated that it is willing to encourage technology fusion beyond national borders. The International Superconductivity Technology Center (ISTEC), established in

1988, was one of the first Japanese consortia to invite foreign participation. Participants are from a wide range of industries that are working with Japanese electronics and automobile companies and include DuPont in the United States. The government is also taking steps to make it easier for foreign firms to join Japan's national projects. Intelligent Manufacturing System, Micromachine, and other new programs were conceived as international projects. MITI's projects with plural foreign participants are listed in Table 5-9, and those with a single foreign participant in Table 5-10.

Japan's experience has relevance for U.S. policymakers. In recent years, U.S. government has increased its support for collaborative research. Examples include SEMATECH, the Department of Commerce's advanced technology program, and the Department of Energy's battery consortium among the Big Three automakers. If the Japanese experience is valid and if technology fusion is facilitated by wide industry membership in collaborative research, the United States may want to focus on bringing a variety of competencies into its consortia.

TABLE 5-9
MITI projects with plural foreign participants, 1992

Name of Project (budget in ¥ million; period)	Total no. of Participants	Foreign Participants
Material Functions Energy Global Environment (676; 1988–1996)	113	48 foreign participants, including: United Technologies Research Center (US) MIT (US) CNRS (France)
R&D for High TC Superconductor (3,750; 1988–)	109	9 foreign participants, including: DuPont (US) IBM Japan
Super/Hyper-Sonic Transport Propulsion System (3,767; 1989–1996)	7	Rolls-Royce (UK) SNECMA (France) United Technologies/Pratt & Whitney (US) General Electric (US)
Micromachine Technology (2,023; 1991–2000)	27	IRS Robotics Corp. (US) SRI International (US) Royal Melbourne Institute of Technology (Australia)
Advanced Telecommunication System (1,708; 1986–1996)	139	IBM Japan DEC Japan Yokokawa Hewlett-Packard
Protein Engineering (1,952; 1986–1996)	14	DEC Japan Roche Japan

TABLE 5-10
MITI projects with a single foreign participant, 1992

Name of Project (budget in ¥ million; period)	Total no. of Participants	Foreign Participant
Advanced Chemical Processing Technology (2,203; 1990–1998)	23	SRI International (US)
High-Performance Materials for Severe Environments (1,670; 1989–1996)	13	Crucible Materials Corp. (US)
Non-Linear Photonics Materials (530; 1989–1998)	9	BASF Aktiengesellschaft (Germany)
Molecular Assemblies for a Functional Protein System (506; 1989–1998)	7	Gesellschaft fuer Biotechnologie Forschung (Germany)
Production and Utilization of Complex Carbohydrates (284; 1991–2000)	6	Pharmacia LKB Biotechnology AB (Sweden)
Quantum Functional Devices (514; 1991–2000)	6	Motorola (US)
New Models for Software Architecture (284; 1990–1997)	2	SRI International (US)
Biological Production of Hydrogen by Environmental Technologies (134; 1991–1998)	6	Eniricerche (Italy)

The U.S. position on foreign participation in government-sponsored collaborative research is unclear at this point. Because technical competencies are found in firms throughout the world, however, the exclusion of foreign nationals from government-sponsored research consortia may limit technology fusion. It should not be assumed that a country can cover the full spectrum of needed technological competence. Foreign companies with unique technical competencies may enhance the probability that global networking will result in heightened technology fusion in fields in which domestic organizations do not have high competence.[36]

CONCLUSION

By contrasting high technology with modern technology, we ascertained that technology fusion is an important characteristic of high-technology innovation. Technology fusion was also contrasted with technical breakthroughs, and a number of revealing comparisons were made between technical innovations achieved through technology fusion and those achieved through technical breakthroughs.

We found that the essence of technology fusion is reciprocal R&D investment and that reciprocal research investment is more than simply an analytical tool for studying technology fusion. Because technology fusion goes beyond a combination of technologies, it is unlikely to be attained through merger and acquisition; rather, it requires joint research among a variety of companies in different industries.

The quantitative analysis clearly suggested that the technical innovations involved in high technology are better described as technology fusion than as technical breakthroughs. Yet one technical breakthrough alone is not sufficient for making progress in high technology. Only through the organic fusion of several technical breakthroughs in a number of fields can a new technology be created.

As we have described the dimensions of techno-paradigm shift from manufacturing through product development, from individual companies to government policies, it has become apparent that technology fusion is implicit in the other dimensions of techno-paradigm shift. The changing focus of manufacturing companies, the emerging trajectory of business diversification, the changing pattern of R&D competition, and the increasing importance of demand articulation are all related and together send a clear message to management: technology fusion is an increasingly important strategy for creating new products and new materials.

In order to realize technology fusion, it is necessary to tap into the "core competencies" of a variety of companies across many industries in different countries. This can be done by forming global partnerships based on concerted efforts to pool complementary assets in an interactive way. Such partnerships may be established among domestic companies or through government-organized research consortia, which may include foreign companies with unique competencies. One cannot assume that the full spectrum of needed competence is present within a country; thus, looking to foreign companies increases the likelihood of achieving technology fusion in areas where domestic companies do not have high competence.

Thus far we have examined the production of technologies. Now we will consider the diffusion of technologies, specifically, how high technologies such as information technology are being used throughout society.

NOTES

1. "Can the Japanese create?" *Newsweek*, 2 July 1984, 27.

2. S. Feiman and W. Fuentevilla, *Indicators of International Trends in Technological Innovation* (Washington, D.C.: National Science Foundation, 1976).

3. F. Narin and J. Frame, "The Growth of Japanese Science and Technology," *Science* 245 (1989): 600–05.

4. F. Kodama, "Interview: The Erosion of American Technological Competitiveness," *Science & Technology in Japan* 9, no. 34 (1990): 4–9.

5. N. Rosenberg, *Inside the Black Box: Technology and Economics* (Cambridge: Cambridge University Press, 1983), p. 55.

6. F. Kodama, "Interview: The Erosion of American Technological Competitiveness."

7. F. Narin and J. Frame, "The Growth of Japanese Science and Technology."

8. Richard Nelson, "Policy Implications of Japan's Growing Technological Capabilities: Framing the Issues," in *Japan's Growing Technological Capability*, ed. T. Arrison et al. (Washington, D.C.: National Academy Press, 1992), 209–15.

9. G. Bylinsky, "The High-Tech Race: Who's Ahead?" *Fortune*, 13 October 1986, 18–36.

10. F. Narin and J. Frame, "The Growth of Japanese Science and Technology."

11. B. Johnstone, "Mechanics Meets Electronics," *New Scientist* 110, no. 1503 (1986): 35–38.

12. N. Rosenberg, *Perspectives on Technology* (Cambridge: Cambridge University Press, 1976), 28.

13. Japan Industrial Policy Research Institute, *A Study of Innovation Process* (in Japanese) (Tokyo: Sangyou-Kenkyusho, 1989).

14. F. Kodama, "Japanese Innovation in Mechatronics Technology: A Study of Technological Fusion," *Science and Public Policy* 13, no. 1 (1986): 44–51. For a broader discussion, see F. Kodama, "Can Empirical Quantitative Study Identify Changes in Technoeconomic Paradigm?" *Science, Technology and Industry Review* no. 7 (1990): 101–29.

15. N. Rosenberg, *Inside the Black Box: Technology and Economics.*

16. N. Rosenberg, *Perspectives on Technology*, 9–31.

17. J. Tilton, *International Diffusion of Technology* (Washington, D.C.: The Brookings Institution, 1971), 49–73.

18. R. Levin, "The Semiconductor Industry," in *Government and Technical Progress: A Cross-Industry Analysis*, ed. R. Nelson (New York: Pergamon Press, 1982), 9–100; Organization for Economic Cooperation and Development, Case Study of Electronics with Particular Reference to the Semiconductor Industry (joint working paper of the Committee for Scientific and Technological Policy and the Industry Committee on Technology and the Structural Adaptation of Industry, Paris, 1977), 133–63; N. Rosenberg, "Civilian 'Spillovers' from Military R&D Spending: The American Experience since World War II," in *Technical Cooperation and International Competitiveness*, ed. H. Fusfeld and R. Nelson (Troy, N.Y.: Rensselaer Polytechnic Institute, 1988), 167–87.

19. F. Kodama, "Interview: The Erosion of American Technological Competitiveness."

20. Ministry of International Trade and Industry, *Electronics & Machinery Industry in the Seventies* (in Japanese) (Tokyo: Tsushou-Sangyou-Chousakai, 1971), 50.

21. Prime Minister's Office of Japan, Statistics Bureau, *Report on the Survey of Research and Development* (in Japanese) (Tokyo: Nihon-Toukei-Kyoukai, 1970–1982).

22. F. Kodama, "Technological Diversification of Japanese Industry," *Science* 233 (1986): 291–96.

23. Johnstone, "Mechanics Meets Electronics," 36–37.

24. "The Search for Superdrugs," *Business Week*, 13 May 1991, 92–96.

25. M. Dibner, "Biotechnology in Pharmaceuticals: The Japanese Challenge," *Science* 229 (1985): 1230.

26. G. Bylinsky, "The High-Tech Race: Who's Ahead?"

27. Private conversation with the author.

28. Ibid.

29. Akio Morita, "Partnering for Competitiveness: The Role of Japanese Business," *Harvard Business Review* 70, no. 3 (1992): 82.

30. Economic Research Institute, *A Survey on the Interaction among Important Indus-*

trial Technologies (in Japanese) (Tokyo: Japan Society for the Promotion of Machinery Industry, May 1990).

31. I. Shirai and F. Kodama, "Quantitative Analysis on Structure of Collective R&D Programs by Private Corporations in Japan," NISTEP Report No. 5 (Tokyo: National Institute of Science and Technology Policy, 1989).

32. C. K. Prahalad and G. Hamel, "The Core Competence of the Corporation," *Harvard Business Review* 68, no. 3 (1990): 79–91.

33. Akio Morita, "Partnering for Competitiveness: The Role of Japanese Business," 76–83.

34. D. Teece, "Profiting from Technological Innovation: Implications for Integration, Collaboration, Licensing and Public Policy," *Research Policy* 15, no. 6 (1986): 285–305.

35. D. Okimoto et al., *U.S.-Japan Strategic Alliances in the Semiconductor Industry: Technology Transfer, Competition and Public Policy* (Washington, D.C.: National Research Council, 1992).

36. F. Kodama, "Japan's Unique Capability to Innovate," in *Japan's Growing Technological Capability*, ed. C. Bergsten et al. (Washington, D.C.: National Academy Press, 1992), 147–64.

6

Societal Diffusion
From Technical Evolution to Institutional Coevolution

The oft-cited dichotomy between incremental and radical innovations, which high-tech development is making obsolete, is less important when it comes to societal diffusion because we have to analyze how a new technology and its social institution *coevolve*.

In this context, besides incremental and radical innovation, C. Freeman has added two more categories: (1) change in the technology system and (2) change in the techno-economic paradigm.[1] While the effects of such innovations as synthetic materials, petrochemicals, and injection molding have gone beyond specific industrial sectors and thus have changed an entire industrial technology system, they are confined to the manufacturing sector. On the other hand, according to Freeman, the effects of innovations that he terms techo-economic paradigm change, such as steam power and electric power, were so far-reaching that they influenced the behavior of the entire economy. The effects of the combination of innovations associated with the electronic computer have been equally far-reaching. Freeman asserts:

Such changes in paradigm make possible a "quantum leap" in potential productivity which, however, is at first realized only in a few leading sectors. In other sectors such gains cannot usually be realized without organizational and social changes of a far-reaching character.[2]

The growing interest in high technology lies in high-tech diffusion beyond the manufacturing sector and even beyond industry. Because it goes beyond industry, we call it "societal diffusion" rather than technology diffusion. There is a surprisingly small amount of empirical and quantitative evidence regarding this diffusion, although its dimensions are mentioned by several authors.[3] In this chapter, therefore, we will focus on discovering both descriptive and quantitative *evidence* about the shift in diffusion process.

The descriptive evidence of techno-paradigm shift in societal diffusion is drawn from two case studies, one in the field of telecommunications and one in welfare. We have chosen these fields because they are subject to heavy regulation, and we were able to examine how changes in regulations interacted with the diffusion of technology. In telecommunications, changes in regulations (institutional changes) often initate the diffusion of technology. The case study presented here confirms Freeman's assertion that institutional inertia deters technology diffusion.

We will also show, however, that the interaction between regulation and technology often works the other way around: the introduction of new technology triggers a chain reaction of institutional change, which, in turn, creates a favorable environment for the rapid diffusion of the technology. These phenomena are most conspicuous in the medical field, as Freeman pointed out. One such example is the diffusion of computerized axial tomography (CAT) scanners in Japan.[4] Here, the relevance of technology to the social institution (in this case, the Japanese national health insurance system) accelerated rather than discouraged acceptance of the new technology. This is in contradiction to Freeman's statement that while technological change per se is often very rapid, there is usually a great deal of inertia regarding change in social institutions.

The heart of the problem posed by societal diffusion appears to lie in the *coevolution* of technology and social institutions. To find quantitative evidence for this, we can investigate the diffusion of technology into local government. Japan has a unique database that has tracked the installation of computers by forty-seven prefectural governments each year since 1963.[5] The conventional explanation of the diffusion process uses the analogy of epidemic phenomenon, and thus its mathematical formulation is a logistic curve. Our regression analysis will reveal that the diffusion process does indeed follow the conventional model. When it comes to the utilization of the installed computers, however, our analysis will reveal that the diffusion process no longer follows the conventional logistic model.

One of the conspicuous characteristics of information technology is *network* externality. As a network's coverage is extended by linking up additional "subscribers," the cost of providing basic services to each user declines.[6] Integration, however, requires a measure of technical compatibility or standardization. To study *institutional coevolution*, therefore, it is necessary to know how and for what purpose the technology is used, because some uses may involve a substantial change in organization, and some may not. Because the Japanese database divides computer use into forty-one categories, including payroll calculation, taxation, planning, and so forth, it is possible to compare the diffusion processes of the same technology in different institutional environments

We can formulate the problem as follows: Is institutional inertia the determinant of high-tech diffusion? More specifically, does an innovation that involves substantial organizational change diffuse more slowly than an innovation that does not need such changes? Our analysis will reveal that there are two levels of dichotomy: one between organizational and technical complexity, the other between technologies applied to old activities and those applied to new activities. The diffusion of computer utilization is slow when it is applied to those tasks for which organizational change is necessary, even if the usage is technically simple. Diffusion is rapid when it is applied to those tasks for which no substantial organizational change is needed, even if usage is technically sophisticated. This clear-cut distinction, however, has one exception. When the computer is applied to new activities, its diffusion is quick, even if it involves organizational complexity, because an institutional framework for new activities *coevolves* with the diffusion of the technology.[7]

We can translate these findings on diffusion dynamics into business practice to discover how a company can respond positively to the technical challenges imposed by abrupt regulatory change. If this change in regulation is caused by societal needs, such as public concerns about environment, a company can implement technological change and profit. We can illustrate these possibilities by describing how the Japanese auto industry managed the transitional process first to comply with the change in emission standards and then to accommodate economic concerns about fuel economy.

DESCRIPTIVE EVIDENCE

The diffusion of fax machines did not begin until regulatory changes had been made. The revision of the Public Telecommunications Law in 1971 and the consequent opening of the telephone network to free use, encouraged the latent demand for facsimile communications. Because the fax is suited to all written languages, the machine's diffusion in Japanese was rapid.[8]

On the other hand, the diffusion of CAT scanners in Japan triggered a chain reaction of institutional change, which further accelerated the diffusion of the technology. The first CAT scanner was developed in the United Kingdom, but its diffusion rate was low in Japan. When the number of CAT scanners surpassed 200 units, however, a professional association of experts was established. This association demanded that the Ministry of Health permit CAT scans to be covered by the national health insurance plan. The government acquiesced, and the market for scanners expanded rapidly.[9]

Using the following two case studies, we will show that one of the key determinants for societal diffusion of high technologies is institutional rather than technical. Then, we will try to identify other critical determinants that affect how technology diffuses within an institutional framework.

Fax Machines

Telephotography was a forerunner of today's fax machines. As early as 1930, the Japanese Ministry of Communications started telephotography service between Tokyo and Osaka. During that period, however, the service was used primarily for the transmission of photographs and diagrams for the news media and weather service.

Subsequently, the police, as well as large banks and electric and railroad companies, which needed to transmit massive amounts of documents, began to use the fax service. In the years up to 1965, business corporations expanded their use of fax services and local municipalities began to use the service for such tasks as census registrations. The spread of the service, however, was limited because terminal equipment was complex and expensive, as were communication fees. Moreover, since the public telecommunications law did not allow the telephone network to be used for purposes other than *voice* transmission, usable lines for facsimile communication were limited to nonswitched lines only, which were leased to subscribers and thus charged at a fixed rate. This situation was identical to the situation in other countries, so facsimile communication was often referred to as a "sleeping giant."

In May of 1971, the Public Telecommunications Law was amended somewhat to allow usage of the telephone network for purposes other than voice transmission. Deregulation of the network enabled facsimile equipment to be connected to subscriber lines. With this change, charges could be levied according to the length of time and distance used for communication, not according to a fixed rated. With this amendment, it became possible for any facsimile machine to communicate freely with another facsimile machine on the public telephone network if the machines could be properly interfaced. With the freeing of the telephone network, facsimile communications began to spread rapidly.

In 1973, a new public telecommunications service was inaugurated. This was a facsimile service. It was connected to the switched telephone network and enabled a subscriber to connect a fax machine to his line. At the same time, communication equipment manufacturers began to sell equipment to telephone subscribers. One company after another, large and small, began to use the service for general communications, i.e., to send documents between offices. Less than two years after deregulation, the number of fax machines installed doubled, from 8,800 units in 1971 to 16,000 units in 1973.

The rapid growth in use was accompanied by rapid progress in scanning and recording technologies. This was due to intensified competition among equipment manufacturers. At the same time, transmission time was reduced through the development of band-width compression techniques, which made facsimile communication less expensive than voice transmission. Furthermore, efforts were made to develop low-cost, high-reliability, compact fax machines for household use. Moreover, studies were begun on a "facsimile communications system," which would be integrated with a stored and a forward-switching network, thus providing versatile service at low communications costs. This system was inaugurated in September 1981.

As of March 1989, it was estimated roughly that the number of fax machines installed in Japan amounted to 24 percent of the number of the total telephone business lines, which is equivalent to a situation in which one fax machine is available for every four business persons, even if we assume that one telephone line is used by one business person. This level is almost saturation as far as the use of fax machines for business purposes is concerned. It is surprising how rapidly this saturation level was reached. As Figure 6-1 illustrates, in less than ten years, annual production in terms of number of units grew by more

FIGURE 6-1
Growth pattern in production of facsimiles

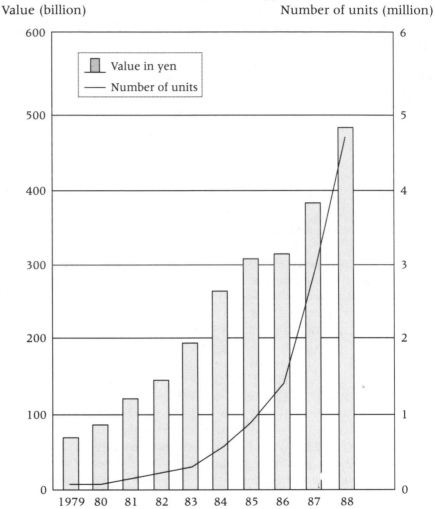

Source: Ministry of International Trade and Industry *Industrial Statistics,* 1989.

than forty times, which indicates that demand for this type of communication was tremendous in Japan.

Although the Japanese language induced early development and diffusion of the fax machine, its diffusion in the rest of the world came later. Because of various technology developments, such as bandwidth compression, made in Japan, facsimile communication surpassed the telex in every functional aspect. Therefore, it diffused rapidly in coun-

FIGURE 6-2
International comparison of installed units of facsimiles

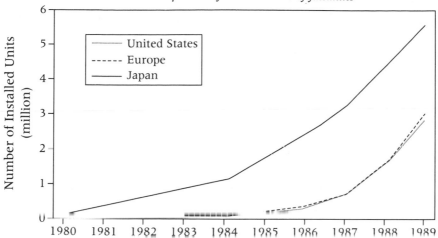

Source: Dataquest for U.S. and European data; The Japanese Association of Communication Equipment for Japanese data.

tries that have different traditions and communicate in different languages. It is interesting to note, as shown in Figure 6-2, that the diffusion paths in United States and Europe are the same in spite of business and cultural differences, which indicates how universal the benefits brought about by fax machines are. In March of 1988, the worldwide installation of fax machines reached 4.3 billion units, and the Japanese share composed only a thin majority—52 percent, as shown in Table 6-1.

CAT Scanners

The computerized axial tomography (CAT) scanner is a highly sophisticated imaging modality that produces cross-sectional pictures of inter-

TABLE 6-1
Number of installed units of facsimiles by region (thousands)

Year	Total	Japan	N. America	Europe	Others
1987	3,250	1,800 (55%)	800 (25%)	400 (12%)	250 (8%)
1988	4,250	2,200 (52%)	1,100 (26%)	600 (14%)	350 (8%)

Source: Nikkei Communications (in Japanese), 28 March 1988, 46.
Notes: World share is in parentheses. 1987 data is units as of September 1987; 1988 data is as of March 1988.

nal organs of the body, using a special configuration of X-rays, detectors, and computers.

Electrical Musical Industries (EMI) Ltd., a British firm, developed high resolution TVs in the 1930s, pioneered airborne radar during World War II, and developed Great Britain's first solid-state computers in 1952. In the late 1960s, Godfrey Houndsfield, an EMI senior research engineer, engaged in pattern recognition research. His research resulted in a scan of a pig's brain. Subsequent clinical work established that CAT technology was viable for generating cross-sectional views of the human body, the greatest advance in radiology since the discovery of X-rays in 1895. With the cooperation of the Department of Health and Social Security, EMI developed the first X-ray CAT scanner.[10]

In Japan, CAT scanners were expensive and therefore diffused slowly. The Japanese government purchased only one unit, which was for the Research Institute of Legal Medicine. However, the Japanese government had been trying to solve the problem of trade surplus since 1976. When it was announced that Queen Elizabeth would visit Japan, the Japanese government decided to purchase thirty-three CAT scanners, at a price of ¥200 million each. The idea for the purchase came from members of the Automotive Insurance Association who were facing increasing claims. At the time, ownership of passenger cars had increased dramatically in Japan, and traffic accidents caused many cases of brain damage, claims for which were expensive for the insurance agencies. Insurance companies needed a reasonably sure way to ascertain whether or not there was brain damage at the time of an accident. The purchased CAT scanners were distributed to brain surgery departments in major public hospitals, giving many doctors their first chance to use the scanners. In medical schools, students learned medicine through cross-sectional pictures of the human body, but in practice, they could only view pictures of vertical section, which were provided by X-ray equipment. After using the scanners, many doctors felt they could not do without them. As a result, Japan became a country with many scanners.

When the number of CAT scanners surpassed 200 units, the number of experts reached 1,000, because 5 experts are needed on average per CAT scanner. At this point, a professional association of experts was established. The association became a lobbying group that demanded the Ministry of Health permit CAT scans to be covered by the national health insurance plan. In November 1978, the government acquiesced, and the market for scanners expanded rapidly. Although Japanese medical equipment manufacturers underestimated the market potential for CAT scanners in the early stages, they soon recognized the profit potential, initiated research and development efforts, and eventually built a mass production system. Thus the average price of a CAT scanner was reduced to one-fourth—¥50 million—of its original price. In 1988, Japanese production of CAT scanners reached ¥76.3 billion, and exports reached ¥33.3 billion.

As of 1988, approximately 6,000 units had been installed in Japanese hospitals, as shown in Table 6-2. There are two types of CAT scanners: a head scanner for the brain only and a body scanner for the whole body. Although early scanners were for the head only, as many as three-quarters of CAT scanners installed by 1988 were body scanners, and 90 percent of both types are made by Japanese manufacturers. In his book *Economic Analysis of Product Innovation: The Case of CT Scanners*, M. Trajtenberg noted that although CAT scanners originated in England, the United States became the prime locus of innovative activity almost immediately, and accounted for a full half of the world market for imaging technologies.[11] In order to show the diffusion rate of CAT scanners in Japan, Table 6-3 shows the figures for Japan compared with corresponding U.S. data reported in Trajtenberg's book. As shown in the table, in all the hospitals with more than 100 beds, the Japanese adoption level is equivalent to 94 percent of the U.S. level. When it comes to a size-by-size comparison, however, there is almost no gap between the two countries. Therefore, the difference in total comes primarily from the difference in hospital size between the two countries. In middle-sized hospitals of 300 to 399 beds, the Japanese rate surpasses the U.S. rate.

The mechanism by which a new technology induces changes in regulation and by which its diffusion is accelerated is even more evident when we consider other changes that have occurred in the Japanese medical system. The same Japanese health insurance plan that brought about the diffusion of CAT scanners also encouraged the diffu-

TABLE 6-2
Number of CAT scanners installed in Japan as of August 1988

	Head	Body	Total
Japanese made (A)	1,403	4,072	5,475
Foreign made	110	511	621
TOTAL (B)	1,513	4,583	6,096
Share (A/B)	93%	89%	90%

Source: New Medicine (in Japanese) (October 1988): 75.

TABLE 6-3
United States-Japan comparison in diffusion rate of CAT scanners, 1984

Number of beds	% of hospitals with CAT		
	Japan (A)	USA (B)	(A/B)
100–199	28.8	29.2	0.98
200–299	62.2	64.0	0.97
300–399	86.1	84.6	1.02
400–499	96.2	97.1	0.99
500–	100.0	100.0	1.00
All	*53.9*	*57.0*	*0.94*

Sources: For U.S. data, M. Trajtenberg, *Economic Analysis of Product Innovation: The Case of CT Scanners* (Cambridge: Harvard University Press, 1990); for Japanese data, Japan Hospital Association, *Reports on Hospitals* (in Japanese) (Tokyo; Byouin-Gaikyou-Chyousa-Houkoku, 1990).
Notes: U.S. data is only up to 1984 and does not include hospitals with less than 100 beds. Moreover, data does not make a distinction between head and body scanners. In the Japanese data, the combined adoption rates of head and of body scanners is calculated, although there is a danger of duplication.

sion of other products including extracorporeal shock wave lithotripters (ESWLs). More ESWLs are used for kidney stone removal in Japan than in any other nation: 1.28 units per million population in 1989 compared with 0.92 in the United States, 0.93 in Germany, and 0.52 in France.[12] This is somewhat surprising as the incidence of kidney stones is not significantly higher in Japan than elsewhere. Moreover, the manufacturer of ESWLs is still dominated by European companies. The 1991 market share of ESWLs installed in Japan by two German and one French suppliers was 64 percent and only 0.5 percent by the largest Japanese supplier.[13] These figures contradict Freeman's argument that there is usually great inertia toward change in social institutions, although technological change per se is often very rapid.

Key Determinants of Diffusion

From these two case studies, we have discovered that one of the key determinants for societal diffusion of high technologies is institutional rather than technical. What, then, are the other determinants within an institutional framework?

According to a sample survey conducted by the Japan Management Association, the fax machine has been more widely adopted than other popular office equipment, such as word processors, high-speed plain

TABLE 6-4

Adoption rate of office equipment, 1988

	Small offices	Large offices
facsimile	96.0%	96.8%
word processor	81.3	93.5
PPC (high speed)	85.7	90.3
personal computer	71.8	95.3

Source: Japan Management Association, *White Paper of Office Automation* (in Japanese) (Tokyo: Japan Management Association, 1990).

Note: Small office is defined either as a company office of 50 to 300 employees or as a local government office of a population of less than 100,000.

paper copiers (PPCs), and personal computers. The survey's results (shown in Table 6-4) indicate that there is no difference between small and large offices in the adoption rate of fax machines, but size does make a difference for other office equipment. A careful look at adoption rates by size, including offices with less than fifty employees, however, reveals that size is indeed a critical determinant of diffusion. As shown in Table 6-5, the adoption rate by offices with less than ten employees surpassed 50 percent only in 1988, when the rate was almost 100 percent in offices with more than 300 employees. Therefore, even if a technology is widely accepted, size is a critical factor in diffusion.

The diffusion path of CAT body scanners by hospital size is shown in Figure 6-3 and confirms our findings. In 1989, the adoption rate reached 100 percent in hospitals with more than 500 beds, while it remained less than 25 percent in those with fewer than 100 beds. And the estimate of the adoption rate in all hospitals reached 53.6 percent in 1989. According to the data, adoption rates had almost reached the saturation point in hospitals with more than 200 beds, while the rate

TABLE 6-5

Adoption rate of facsimiles by office size as of 1988

Number of Employees	Percentage of Adoption
5–9	55.2
10–19	73.1
20–49	83.3
50–99	90.8
100–299	92.7
–300	99.0

Source: Data Communication (in Japanese) (May 1990): 43.

FIGURE 6-3

Diffusion path of body CAT scanners by hospital size

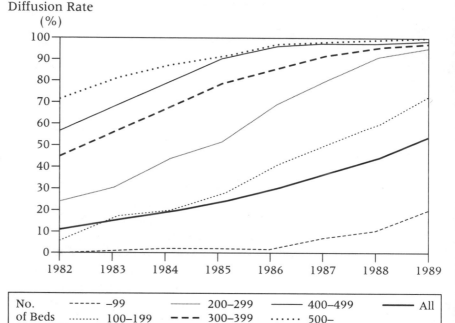

Source: Number of hospitals, Ministry of Health; diffusion of CAT scanners, Japan Hospital Association, *Reports on Hospitals* (in Japanese) (Tokyo: Byouin-Gaikyou-Chyousa-Houkoku, 1990); diffusion data for 1988, The Japan Hospitals Federation, *Management Analysis Survey of Hospitals in Japan* (in Japanese) (Tokyo; Byouin-Keieibunseki-Chyousa-Houkoku, 1989).

Note: A hospital is defined as a medical institution that has the capacity for more than twenty patients. However, hospitals for the treatment of psychiatric problems, tuberculosis, and leprosy are not included in this figure.

for hospitals with fewer than 200 beds was still growing. Therefore, the time path of the adoption rate in all the hospitals is still following the path of exponential growth and size was a critical determinant.

QUANTITATIVE EVIDENCE

In the past twenty years, many economists have worked to understand the process involved in the diffusion of technologies. Among these economists, Edwin Mansfield has carried out an extensive econometric analysis of modern time-series samples. He traced the extent of adoption for many specific production technologies within firms and industries.[14] We are interested, however, in the process of technology diffusion beyond a few leading sectors of high technologies into the entire society i.e., societal diffusion. To study this, we need to focus on those sectors that used to have little to do with exploiting the potentials of new technologies because high-tech development is now making exploitation possible and effective for these sectors. The diffusion of computers in local governments is one example of high-tech development that can be exploited by sectors not usually involved in new technologies.

Conventional Model

According to Paul David, Mansfield's point of departure for his work was a model of diffusion involving a new technology with prespecified engineering and economic characteristics, an *unchanging* population of potential users, and an objective economic environment in which the only consequential change was the gradual dissemination of information. In Mansfield's view, the gradual increase in the extent of an innovation's application across firms and sectors of the economy is an adjustment process that eventually restores the equilibrium in saturation.[15]

Sociologists have studied the nature and sources of information that managers obtain concerning new techniques.[16] The sources of information vary, depending on how close the manager is to adopting the innovation. In agriculture, for example, mass media is the most important source at the early stages of a manager's awareness of the innovation, but friends and neighbors are most important sources when a manager is ready to try the innovation. Based on this observation, sociologists have developed the theory of a "two-step flow of communication." The early users of an innovation tend to rely on sources of information beyond their peer group's experience; after they have begun using the innovation, they become models for their less expert peers, who can intimate their performance.[17]

A number of other authors have also made important contributions to the literature on diffusion. The assumption behind all this literature is, however, that information-spreading mechanisms are based on personal contact.[18] According to Stoneman, the Mansfield model is a psychological model in the sense that the speed of response is assumed to be related to the size of the stimulus. It is also epidemic in the sense of learning by infection.[19] In this psychological approach, therefore, the assumption about an agent's decision to adopt a technology is that uncertainty is reduced in proportion to the number of agents already using the technology. Therefore, a logistic curve is obtained to suggest the diffusion pattern.

The basic hypothesis of the simple epidemic model is represented by the following equation:

$$m_{t+1} - m_t = \beta \cdot (n - m_t) \cdot m_t, \beta > 0,$$

where m_t is the number of individuals in a fixed population of n having contracted an infectious disease at time t. Thus, the number of individuals contracting the disease between times t and $t + 1$ is proportionate to the product of the number of uninfected individuals and the proportion of the population already infected, both at time t. If the period t to $t + 1$ is small, it may be alternatively stated as

$$dm_t/dt = \beta \cdot (n - m_t) \cdot m_t.$$

This differential equation has the following solution:

$$m_t/n = \{1 + \exp(\alpha - \beta \cdot t)\}^{-1}, \tag{1}$$

where α is a constant of integration.

New Dimensions

According to Paul David, an important dimension in the diffusion of high technologies is the presence of network externalities.[20] These are most conspicuous in the field of telecommunications, where decisions regarding terminal equipment are affected by the costs of access to other parties over existing networks. Certainly, this dimension is important in any study of the coevolution of technology and social institution.

In the case of computers, hardware costs are only a part of the story since the available range and price of compatible software are also important in determining the use that will be made of the technology. Although network technologies exhibit some of the same dynamic features in their development as "freestanding" product and process innovations, they pose different economic problems. Integration requires some measure of technical compatibility or standardization, thus making the systems public goods. Therefore, the benefits derived by any one prospective user may be dependent on the willingness of others to incur the costs of achieving compatibility with the same network. Where there are alternative, emerging network technologies to choose from, the public goods problem tends to slow the adoption of any one technology. Potential subscribers will wait in the hope that others will bear the costs of compatibility.

Once we have moved beyond a static conception of technologies, however, there is another important point to be made about the dynamics of technology choices: the existence of irreversible, dynamic-scale economies means that small, initial advantages can readily cumulate into larger ones. This opens up the possibility that a particular systems design can become "locked in," and rival technologies can be "locked out" in a competitive market.[21] Therefore, network externality may first slow and then accelerate the diffusion of technology.

The distinction between diffusion and adoption of technology is another important dimension in a study on *coevolution*. Metcalfe makes this distinction.[22] Diffusion analysis, according to Metcalfe, is concerned with the economic significance of a new technology. The market share of a product innovation or the fraction of industry output produced with a process innovation is often used to measure diffusion. In this sense, the analysis of diffusion is closely related to the analysis of technological substitution. Adoption analysis, in contrast, considers agents' decisions to incorporate a new technology into their activities. It leads to propositions linking the nature and timing of adoption decisions to specified characteristics of adopters. Our interest in societal diffusion lies in the process by which new technology is adopted for the first time by sectors that have never used the technology. Because this is obviously more than technological substitution, we need adoption analysis rather than diffusion analysis.

Database

Adoption analysis requires the following: (1) an *unchanging* population of potential adopters; (2) *homogeneity* in missions among adopting agents; (3) *independence* in decision making among agents; and (4) *heterogeneity* of adopter characteristics. We can find a fairly unique Japanese database that accommodates these requirements.

The Japanese Ministry of Home Affairs counts the number of computers installed by local governments. For the past 100 years, Japan has been divided into forty-seven administrative units, or prefectures; therefore, there is an unchanging population of potential adopters. Prefectural governments are homogeneous by nature and similar in organizational principle and missions. In fact, their responsibilities are specified in detail by law. They are, however, heterogeneous in terms of size and decision making. Because the division into prefectures is based on land area, not population size, the prefectures are different in terms of size as measured by population as well as by budgets. While the budgets of the central government's ministries and agencies are directly controlled by the Ministry of Finance through budget allocation, there is no similar control over prefectural governments. Therefore, it can be assumed that they are independent decision makers. The dates of computer installation are shown in Table 6-6. The real advantage to this database, however, is that it accommodates some organizational aspects of technology diffusion. The survey asks whether or not the installed computer is used for each of forty-one task categories. In other words, we can identify when installed computers were first used to perform each of the forty-one specified task categories for every year since 1963.

Out of the forty-one task categories, there are nine in which the diffusion rate is relatively high by the year 1985. However, there are some categories in which the available data seem to be less reliable because of an obvious inconsistency: the diffusion rate peaked and

TABLE 6-6
Diffusion of computer installation in prefectural governments

Year	Number of Prefectures	Diffusion Rate
1963	1	0.02
1964	2	0.04
1965	4	0.09
1966	5	0.11
1967	7	0.15
1968	12	0.26
1969	14	0.30
1970	20	0.43
1971	23	0.49
1972	27	0.57
1973	39	0.83
1974	43	0.91
1975	44	0.94
1976	45	0.96
1977	46	0.98
1978	47	1.00
1979-	47	1.00

Source: Management and Coordination Agency, *Annual Report on Utilization of Computers in Government* (in Japanese) (Tokyo: Nihon-Toukei-Kyoukai, 1985).
Note: The definition of a computer is in accordance with that established by MITI: it should be digital processing and programmable and have more than 2,000 bits memory and electronic logical calculation.

decreased drastically afterward. On the basis of the data reliability, therefore, the following five task categories were selected for the statistical test: personnel management; management of issuing bonds; automobile taxation; forecasting and planning; and cost estimation of civil works. Time-series data of the diffusion rate of computer utilization for these five categories are shown in Table 6-7.

We have noted that network externality is one of the new dimensions for a societal diffusion study. Based on the extent to which network externality is relevant, therefore, we will break these five categories into two groups. The first group is composed of personnel management; management of issuing bonds; and automobile taxation. These tasks are characterized by routine work and mechanistic processing but involve large-scale data processing, operation in a central-

TABLE 6 7
Diffusion rates of computer utilization for task categories

Year	Personnel Management	Management of Issuing Bonds	Automobile Taxation	Forecasting & Planning	Cost Estimate for Civil Works
1963	0.04	0.02	0.02	0	0
1964	0.04	0.02	0.02	0	0
1965	0.04	0.02	0.04	0	0
1966	0.04	0.02	0.04	0	0
1967	0.04	0.02	0.07	0	0
1968	0.09	0.02	0.15	0.02	0
1969	0.13	0.02	0.24	0.09	0
1970	0.26	0.02	0.33	0.13	0
1971	0.33	0.09	0.39	0.13	0.04
1972	0.46	0.20	0.46	0.24	0.04
1973	—	0.21	0.55	0.30	0.04
1974	—	—	0.62	0.43	—
1975	—	—	0.62	0.51	—
1976	0.81	—	—	0.66	0.51
1977	0.77	0.57	0.75	0.55	0.45
1978	0.79	0.66	0.72	0.60	0.51
1979	0.79	0.64	0.75	0.64	0.60
1980	0.79	0.72	0.75	0.68	0.62
1981	0.79	0.77	0.77	0.70	0.64
1982	0.85	0.79	0.79	0.70	0.77
1983	0.85	0.79	0.81	0.75	0.85
1984	0.85	0.81	0.79	0.70	0.77
1985	0.85	0.83	0.81	0.70	0.83

Source: Ministry for Home Affairs, *Annual Report on Computers in Local Governments* (in Japanese) (Tokyo: Government Printing Office, 1986).
Note:—, not available.

ized mode, interdependence on other agents, reliance on communication networks, and a high level of standardization. Since network externality matters to this group, we will call these categories of computer utilization "network mode" and label them the "N-mode" group.

The second group is composed of the last two categories: forecasting and planning and cost estimation for civil works. These categories are characterized by unstructured work and organic processing but involve only small-scale, decentralized, independent, operation in a local-use mode, and less standardization. Since network externality is less relevant to this group, we will call these categories "local-use mode," and label them the "L-mode" group.

Installation versus Utilization

When a diffusion path follows the logistic curve, we can assume that the determinant of diffusion is the speed of information-spreading based on personal contact. For the logistic curve of equation (1), empirical tests are fairly straightforward and use regression analysis on the linear transformation:

$$\log\{m_t/(n - m_t)\} = \alpha + \beta \cdot t.$$

Applying the regression analysis to the time-series data of computer installation yields the following results with the coefficient of determination R^2 and t-value in parentheses:

$$\log\{m_t/(n - m_t)\} = -4.3281 + 0.5279 \cdot t, \qquad R^2 = 0.986,$$
$$(-26.91) \quad (29.84)$$

As can be seen, the R^2 is extremely high, and it has a higher t-value. On this basis, we can identify the logistic curve as the diffusion path that best describes the diffusion of the installation of computers. Therefore, the installation behavior is, in fact, psychological and epidemic. In other words, the decision to install the computers is based on psychological and epidemic pressures with less institutional consideration. Therefore, the speed of the diffusion is determined by the speed of information spreading.

To test the evidence of institutional coevolution, the logistic curve was applied to the data set of the computer *utilization* for the five task categories previously identified. In this case, the coefficients of determination become lower throughout the five task categories, as shown in Table 6-8. This indicates that the diffusion mechanism of utilization of installed computers is not based on information spreading or technical adjustment. As shown in the table, none of coefficients of determination is above 0.9, while the coefficient of determination was 0.986 in computer installations. This indicates that the diffusion of computer utilization follows a different path from that followed by computer installation and suggests that diffusion of computer utilization is based on institutional factors.

A closer look at this table will bring to light another distinction. The group with a coefficient of determination above 0.8 and the group with a coefficient below 0.8 seem to coincide with the distinction made between network-mode and local-mode with one obvious exception. Although "automobile taxation" has been identified as part of the N mode group, its coefficient of determination is 0.774, far below 0.8.

This result reveals the unique feature of this task category within the N-mode group. We have already noted such implications of network externality as the "locked in" phenomenon and its retardation effect, which is derived from its public goods nature. Therefore, we need to examine the growth pattern of demand for each task category in the N-mode group. For this purpose, we have used the following indicators: number of employees in prefectural government for personnel management; the prefectural budget for bond issuing; and the number of passenger cars subject to taxation. These growth indices from 1963 to 1985 are shown in Table 6-9.

As the table shows, registration of cars in Japan increased rapidly during the 1970s when compared to growth in other task categories.

TABLE 6-8
Test of logistic curve

Task Category	Mode	Coefficient of Determination	t-value
Computer Utilization			
personnel management	(N)	0.858	(10.44)
issuing bonds	(N)	0.898	(12.61)
automobile taxation	(N)	0.774	(8.27)
forecasting & planning	(L)	0.713	(6.30
civil works' cost estimate	(L)	0.795	(6.52)
Computer Installation		0.986	(29.84)

TABLE 6-9
Growth index of car registration compared to other growth indices
(Index: 1963 = 1.00)

Year	Number of Employees in Prefectural Governments	Budget Size in Prefectural Governments	Number of Passenger Cars Registered
1963	1.00	1.00	1.00
1965	1.11	1.28	1.74
1970	1.19	2.52	6.28
1975	1.31	3.89	13.74
1980	1.41	5.04	19.97
1985	1.44	6.27	23.96

Source: Prime Minister's Office, Statistical Bureau, *Japanese Statistics* (in Japanese) (Tokyo: Nihon-Toukei-Kyoukai, 1986).

In fact, the growth index for car registration increased almost twenty-five times during this period, while the growth index for personnel management experienced almost no increase and that for bond issuing increased only six times. Obviously, the demand for automobile taxation expanded substantially compared to that of other task categories.

The utilization of computers for automobile taxation, therefore, should be analyzed in the context of its extremely rapid growth in demand, rather than in the context of the "locked-in" and "locked-out" effects of network externality. This problem will be mentioned more explicitly and more systematically in the following sections.

INSTITUTIONAL COEVOLUTION

Because of the economists' search for a deeper understanding of the process involved in the diffusion of technologies, a new "microeconomics of innovation diffusion" is emerging. It is based on both theoretical insights and an accumulating body of empirical evidence. As an alternative to epidemic models, it considers the evidence that many new technologies are introduced initially in forms and under market conditions that make them appear profitable in immediate applications for only some firms within the relevant industry. Within those firms, these technologies are profitable perhaps only in the operations of some plants and departments. However, as the new technology and its microeconomic environment coevolve, the extent of profitable application will broaden.[23]

David, Davies, and Stoneman's contributions to the analysis have led to the elaboration of a class of so-called "equilibrium diffusion models." These models emphasize two fundamental points, which can be added to those of Mansfield. The first is that even if information relevant to rational decision making about an innovation were *instantly* disseminated without cost, there would still be many reasons to expect that states of equilibrium would exist involving less than complete diffusion of a new technology within an industry. Second, one should note the dynamic forces that give rise to a "moving equilibrium" in the potential level of fully informed adoption of innovations by rational, profit-seeking agents.[24]

In mathematical terms, the new approach is called the "probit" model. While the conventional epidemic model is based on learning by infection, the probit approach assumes that not all firms use the new technology instantaneously because it is not profitable for all firms to do so.[25] The probit approach is based on an explicit behavioral assumption: whenever or wherever a stimulus variate takes on a value exceeding a *critical* level, the subject of the stimulation responds by instantly deciding to adopt the innovation in question. These decisions are not arrived at simultaneously by the entire population of potential adopters because at any given point in time either the "stimulus variate" or "critical level" required to elicit an adoption is described by a *distribution* of values, not a unique value appropriate to all members of the population. Through an exogenous or indigenous process, however, the relative position of stimulus variate and critical response level are altered as time passes and bring a growing proportion of the population across the threshold into the group of actual users of the innovation.[26]

Size Distribution

Because the probit model requires us to identify a critical variate and its distribution curve, we need to identify these two items for computer diffusion into local governments.

We know from the case studies on CAT scanners and fax machines that size is a critical determinant of diffusion. In the case of fax machines, the adoption rate by offices of less than 10 employees surpassed 50 percent only in 1988, although the rate was almost 100 percent in offices of more than 300 employees. The 1989 adoption rate of body CAT scanners reached 100 percent in hospitals with more than 500 beds, while it remained less than 25 percent in those with fewer than 100 beds.

TABLE 6-10
Statistical test for size distribution

Year	χ^2-value
1960	4.11
1965	5.24
1970	6.51
1975	7.33
1980	5.43
5% point	9.49

Based on these observations, we can reason that the size of a local government constitutes a critical variate in the diffusion of computers. The population is selected as the indicator of size because other indicators, such as budget, are highly correlated to population size.[27]

Now we are interested in identifying the size distribution curve of local governments. We can try to make a statistical test to determine whether or not the population of the prefectural governments is *lognormally* distributed. The data on population distribution in the years of 1960, 1965, 1970, 1975, and 1980 were collected and tested. The distribution curves were obtained by dividing the logarithmic transformation of population into five equal intervals. Then, the χ-*square* test was applied to these data. The results of the test are shown in Table 6-10. Because we are interested only in the fitness of the distribution curve, the degree of freedom is 4. Since the 5 percent point of χ^2-distribution on 4 degrees of freedom is 9.49, as shown in the table, the null hypothesis is not rejected in all the years. That is, one cannot deny that the population distribution follows the lognormal distribution. On the basis of these results, we can assume that the size distribution of the prefectural governments does follow the lognormal distribution.

Davies' Theory

Having learned that size is a critical variate and that it is lognormally distributed, we will choose from existing models in the probit approach the mathematical model that deals with size in the most explicit way.

Davies builds his diffusion model on the assumption that an agent will use a new technology if the expected payoff period become smaller than a critical payoff period.[28] The expected payoff period is a function of agent size and time. The critical payoff period is also related to agent size. Let $ER_i(t)$ be *agent i's* expectation of the payoff period associated with adoption at time t, and $R_i^*(t)$ be *agent i's* critical payoff period against which adoption is assessed at *time t*, then, *agent i* will adopt the innovation by the time t if

$$ER_i(t) \leq R_i^*(t).$$

The expected payoff period, ER, is a function of agent size S_i, time factor $\theta_1(t)$, and other characteristics $\in_i^1$: $ER_i(t) = \theta_1(t) \cdot S_i^\delta \cdot \in_i^1$, where $\in_i^1 > 0$, $\theta_1(t) > 0$. The critical payoff period, R^*, is also related to agent size, the time factor, and other characteristics: $R_i^*(t) = \theta_2(t) \cdot S_i^\tau \cdot \in_i^2$, where $\in_i^2 > 0$, $\theta_2(t) > 0$. The ownership condition can be rewritten: $ER_i(t)/R_i^*(t) \leq 1$. Since both the expected payoff period and the critical period vary with agent size, the ownership condition can be rewritten again as:

$$ER_i(t)/R_i^*(t) = [\theta(t) \cdot S_i^\beta \cdot \in_i]^{-1} \leq 1.$$

where, $\beta = \tau - \delta$, $\in_i = \in_i^2/\in_i^1 > 0$, and $\theta(t) = \theta_2(t)/\theta_1(t) > 0$.

Davies then makes several assumptions concerning the mathematical features of $\theta_1(t)$, $\theta_2(t)$, and β, the details of which we will describe in the next section. Once these assumptions are made, the ownership condition may be reexpressed as:

$$S_i \geq S_i^*(t),$$

where $S_i^*(t)$ is referred to as critical agent size and $S_i^*(t) = [\theta(t) \cdot \in_i]^{-1/\beta}$. In order to derive the diffusion path, we have to assume a specific functional form in the distribution of agent size. In the case of local governments, we have identified the distribution of agent size as following lognormal distribution. In other words, once the distribution curve is identified, we can develop a theory of diffusion. Let $Q(t)$ be the probability that an agent chosen at random will have adopted an innovation by time t, then

$$Q(t) = \text{Probability } [S \geq S^*(t)].$$

Distinction for Coevolution

We are still interested in obtaining evidence of institutional coevolution since the barriers to the diffusion of computer technology lie more in institutional inertia than in the system's technical evolution. We have focused on network externality as the primary necessity of institutional coevolution. In terms of the computer utilization, therefore, the demarcation between organizational and technical complexity of task categories is critical.

In order to accentuate the importance of organizational factors, however, there should be two groups: one in which the computerization of a task category is technically simple but organizationally complex, and the other in which computerization is technically complex but organizationally simple. In terms of organizational complexity, we have already made a tentative distinction between network-mode (N-mode group) and local-mode (L-mode group).

As far as technical complexity is concerned, the computerization of the L-mode task category goes along with the development of a new science, such as planning science, and requires sophisticated software while the computerization of the N-mode group can be accomplished by ironing out technicalities related to data processing. Moreover, the utilization of the computer for the L-mode group requires sector-specific expertise, while the N-mode group primarily involves using the system for clerical work. Therefore, we can conclude that the L-mode group is technically sophisticated and the N-mode group is technically simple. On this basis, we can note that the introduction of computers into N-mode tasks will require changes that are complex organizationally but simple technically and that the L-mode group includes changes that are sophisticated technically but simple organizationally.

Davies' probit model makes the following distinction: *Group A* innovations are relatively cheap and simple; *Group B* innovations are expensive and technically complex.[29] However, technical simplicity is not necessarily correlated to organizational simplicity because substantial organizational change may be needed to exploit a new technology, even if it is technically less sophisticated. In fact, in the Davies mathematical formulation, the distinction does not necessarily come from technical complexity but from whether the major improvements are concentrated in the early years of the diffusion period or scattered throughout. In the Davies' formulation, improvements may be either technical or organizational. That is, they may come from organizational and institutional changes to accommodate the potential of the new innovation.

We can interpret the Davies' formulation in the following way. The agent's expectation of the payoff period, $ER(t)$, will decline with time because there will be successive improvements in the benefit/cost ratio.

Because of organizational inertia involved in adopting a new technology, the benefit is not fully exploited without organizational change, and the cost of labor displacement, which is incurred by adopting the technology, may be substantial. Innovation suppliers can gradually learn how to lessen institutional inertia by adding a certain function to the new technology system. Most agents' views of the profitability of adoption can be improved by information from the search for how to implement the institutional changes associated with the adoption. Based on these interpretations, we can assume that *ER* will decline monotonously with time, as Davies does. Therefore, $\{d\theta_1(t)/dt\}/\theta_1 < 0$ for *all* t. In Davies' model, however, it is assumed that the rate of decline in *ER* will differ between the two groups as follows:

$$d[(d\theta_1/dt)/\theta_1]/dt \begin{bmatrix} > 0 & \text{for Group A innovations,} \\ = 0 & \text{for Group B innovations,} \end{bmatrix}$$

because, for Group A innovations, there is a slowdown in learning by manufacturers over time and a slowdown in the returns from the information search by potential adopters but there is no such slowdown for the more expensive and complex Group B innovations.

We can interpret this distinction in the following way. For L-mode innovations, the rate of decline in *ER* is attributed to a slowdown over time in the availability of easier technical solutions to accommodate institutional inertia and in the returns from information searches concerning easier organizational solutions. For N-mode innovations, however, such a slowdown is not plausible because easier technical or organizational solutions cannot be found and are not effective.

The agent's critical payoff period, $R^*(t)$, will increase with time because of a lessening in the risk involved in implementing necessary organizational changes and the effect of competitive pressures, e.g., public criticism about not implementing changes. Thus, following Davies, we can assume that R^* will increase monotonously with time. Therefore, $\{d\theta_2(t)/dt\}/\theta_2 > 0$ for *all* t. In Davies' model, however, the general shape of the time path for θ_2 was supposed to vary between Group A and Group B innovations

$$d[(d\theta_2/dt)/\theta_2]/dt \begin{bmatrix} < 0 & \text{for Group A innovations,} \\ = 0 & \text{for Group B innovations,} \end{bmatrix}$$

because it was assumed that the relaxation in R^* will level off over time for Group A innovations but will be more sustained for Group B innovations. However, we can interpret this distinction as the relaxation in R^* will level off over time for L-mode innovations because of the relatively lower risks involved in organizational changes, while it will be more sustained for N-mode innovations because of the higher risks involved.

Combining the conditions imposed on θ_1 and on θ_2, Davies obtained the following relationships:

$$d[(d\theta/dt)/\theta]/dt \begin{cases} < 0 & \text{for Group A innovations,} \\ = 0 & \text{for Group B innovations.} \end{cases}$$

Among the possible mathematical functions that satisfy these conditions, Davies specifically considers the following two types of functions:[30]

$$\theta(t) \begin{cases} = \alpha \cdot t^\phi & \text{for Group A innovations,} \\ = \alpha \cdot e^{\phi \cdot t} & \text{for Group B innovations,} \end{cases}$$

where $\alpha > 0$; $0 < \phi < 1$. The multiplicative form of the central limit theorem suggests that $\in$, representing characteristics other than size and time factors, will be lognormally distributed across agents: $\log \in$ is normally distributed across agents with mean zero and variance σ^2. We have shown that agent size of computer utilization, S, is lognormally distributed: $\log S$ is normally distributed with mean μ_s and variance σ_S^2.

Letting $N(\cdot|\cdot, \cdot)$ be the normal distribution function, Davies derives the following formula for the diffusion path:[31]

$$Q(t) \begin{cases} = N(\log t | \mu_d, \sigma_d^2), & \text{for Group A;} \quad (2) \\ = N(t | \mu_d, \sigma_d^2), & \text{for Group B,} \quad (3) \end{cases}$$

where, $\mu_d = -(\log \alpha + \beta \cdot \mu_s)/\phi$, and $\sigma_d^2 = (\sigma_2 + \beta^2 \cdot \sigma_S^2)/\phi^2$. Thus, there are *two* types of diffusion paths: a *cumulative lognormal* time path for Group A innovations and a *cumulative normal* time path for Group B innovations.

In Group A, the portion of agents using a new technology will follow a cumulative lognormal time path because these innovations experience major improvements in the early years but fewer thereafter. The number of agents using Group B innovations, in contrast, follow a cumulative normal time path because the innovations are improved

FIGURE 6-4
Two patterns of a diffusion path

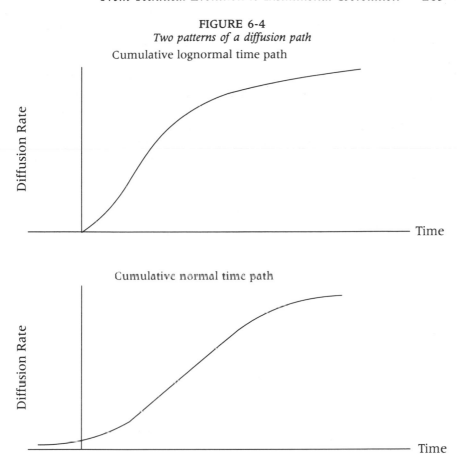

Cumulative lognormal time path

Cumulative normal time path

for many years after their first commercial introduction. The difference in time shape between the paths can be seen in Figure 6-4. As shown in the figure, the cumulative normal time path exhibits a symmetrical S-shaped diffusion curve.[32] The cumulative lognormal time path, on the other hand, takes on a positively skewed diffusion curve.

IDENTIFYING THE DIFFUSION PATH

For an analysis of institutional coevolution, we can interpret the Davies' model in the following way: if the innovation requires a full-scale institutional *reorganization*, it follows the cumulative normal time path (Group B) because major improvements in productivity are possible only after a period of organizational change. If the innovation does not need organizational change, it will follow the cumulative lognormal time path (Group A) because its potential can be exploited easily in the early years.

This brings us to the following hypothesis: N-mode computer utilization follows the cumulative normal time path, while L-mode computer utilization follows the cumulative lognormal time path. To conduct statistical tests for this hypothesis, we have to develop a method for statistical fitting. Both forms of equations (2) and (3) can be linearized by rearranging these two equations in terms of the standard normal function:

$$z_t \begin{cases} = (\log t - \mu_d)/\sigma_d & \text{for cumulative lognormal,} \\ = (t - \mu_d)/\sigma_d & \text{for cumulative normal.} \end{cases}$$

It follows that the diffusion path, $Q(t)$, can be written as:

$$Q(t) = N(z_t | 0, 1).$$

Once these rearrangements are done, given values for $Q(t)$, z_t can be read off from standard normal tables. Therefore, we can estimate the coefficients from the following regression:

$$z_t \begin{cases} = a_1 + b_1 \cdot \log t & \text{for cumulative lognormal,} \\ = a_2 + b_2 \cdot t & \text{for cumulative normal,} \end{cases}$$

where z_t refers to the normal equivalent *deviate* of the level of diffusion (m_t/n) in year t. The regression estimates of a and b can be interpreted as: $a = -\mu_d/\sigma_d$, and $b = 1/\sigma_d$.

Test Results

We can now apply regression analysis to the diffusion of the computer utilization for the five specified task categories. For example, regression analysis applied to the task category of "personnel management," yields the following results with the coefficient of determination R^2 and t-values in parentheses:

(a) for lognormal time path,

$$z_t = -2.850 + 1.216 \cdot \log t, \qquad R^2 = 0.844,$$
$$(-7.626) \quad (8.040)$$

(b) for normal time path,

$$z_t = -2.016 + 0.154 \cdot t, \qquad R^2 = 0.926.$$
$$(-14.04) \quad (14.98)$$

The result shows that all the t-value tests cleared the significance test. That is, the null hypothesis concerning all the parameters can be rejected at a significance level of 0.05. However, the coefficient of determination is higher in the cumulative normal time path than in the cumulative lognormal one, and it is above 0.9 in the cumulative normal time path. Therefore, we can conclude that the cumulative normal time path is the diffusion path of computer utilization for personnel management.

For the other four categories, test results show that there is at least one time path in which the coefficient of determination is above 0.9, and all the t-value tests cleared the significance tests at the level of 0.05. Our purpose then is to identify the diffusion pattern for each of the four task categories. To do so, we can compare the two types of time function in terms of R^2 and its t-value. The results are shown in Table 6-11 with t-values in parentheses. First, in all five categories, the higher coefficient of determination has substantially improved compared to the values that were obtained by fitting the logistic curve (see Table 6-8). Therefore, we can argue that decisions about the utilization of computers are based on expectations of and speculations about the payoff periods, not on psychological and epidemic behavior.

As far as the utilization is concerned, the identification problem centers on a comparison of the cumulative lognormal path and the cumulative normal path. As depicted in the table, there is a fairly clear-cut distinction between the two groups of task categories. Those in the

TABLE 6-11
Comparison of the two time paths

Task Category	Cumulative lognormal	Cumulative normal
Network-Mode		
personnel management	0.844 (9.87)	**0.926** (14.98)
management of issuing bonds	0.740 (7.15)	**0.941** (16.96)
automobile taxation	**0.927** (15.90)	0.900 (13.40)
Local-Mode		
forecasting & planning	**0.938** (15.51)	0.843 (9.27)
cost estimation of civil works	**0.940** (13.17)	0.905 (10.25)

N-mode group are identified as the cumulative normal time path, and those in the L-mode group are identified as the cumulative lognormal path. There is, however, one obvious exception: automobile taxation follows the cumulative lognormal path, although it was assumed to belong to the N-mode group.

The statistical test results described above can be summarized into a taxonomy of computer diffusion (see Table 6-12).

Evidence for Coevolution

Our comparison of the N-mode group and the L-mode group yields the following conclusion: diffusion of computer utilization is slow and follows the cumulative normal time path when it is among the N-mode group of task categories in which organizational change is necessary, even if utilization is technically simple. On the other hand, diffusion is rapid and follows the cumulative lognormal time path when it is among the L-mode group in which no substantial organizational change is needed, even if utilization is technically sophisticated. Thus, the diffusion of computer utilization is, in fact, a function of institutional rather than technical evolution.

A look at two other task categories not previously mentioned—corporate taxation and production of the designated statistics—may reinforce our conclusion if they exhibit strong network externality. The performance of these two tasks is not confined to the area of a prefecture because corporations operate throughout the country and designated statistics have to be aggregated into national statistics. Therefore,

TABLE 6-12
Derived taxonomy of computer diffusion path

	Time Path	Determinant	Examples
Installation	*Logistic*	*Information Spreading*	
Network-Mode Utilization	Normal	Institutional Evolution	Personnel Management Corporate Tax Bonds Issuing Statistics Production
Local-Mode Utilization	Lognormal	Technical Evolution	Forecast & Planning Civil Work Estimate
Network-Mode Utilization	Lognormal	Coevolution	Automobile Tax

the necessity for compatibility in computer utilization among prefectural governments is stronger than in the N-mode categories listed in Table 6-11, which may require compatibility only among offices in the same prefecture. To confirm this hypothesis, we made the same regression analysis of these two categories and obtained the results depicted in Table 6-13. As we can see, the coefficient of determination for the cumulative normal path is extremely high in both categories, far above 0.95, in contrast to the coefficients for other categories, which are below 0.95. Therefore, we can argue that the public goods problems of network externality decelerated diffusion in the early period of introduction and that "locked-in" effects accelerated diffusion in the later period. The stronger the network externality, the more critical is the effect of institutional coevolution in the diffusion of computer utilization.

Now, let us consider a unique feature of the task category of automobile taxation, which was identified as following the cumulative lognormal path, although it was assumed to belong to the N-mode group. Indeed, this category should be identified as one of the highly N-mode categories because the use of cars crosses prefectural boundaries. As shown in Table 6-9, however, car registration increased rapidly during the 1970s. In an expanding area, therefore, we can argue that the introduction of computers does not threaten anyone with unemployment, and if employees' workloads grow because hiring does not keep up with increased work, computers will make workers' lives easier and reduce workloads. Thus, there will be relatively less institutional inertia against the diffusion of computers.

We can generalize this finding to state that within the N-mode group, there are certain tasks that have so expanded that there is no well-established institutional framework for them. The time path for the utilization of computers for these tasks follows the cumulative lognormal one, representing a rapid diffusion pattern. Those task categories for which there is an established institutional framework follow the cumulative normal time path, representing a slow diffusion pattern.

Institutional inertia, then, will be a determinant of diffusion when

TABLE 6-13
Identification of highly N-mode utilization

Task Category	Cumulative Lognormal	Cumulative Normal
Corporate taxation	0.929 (15.39)	**0.990** (41.50)
Production of statistics	0.903 (12.92)	**0.971** (24.51)

the technology is applied to old activities for which the institutional framework is already well established; institutional inertia will not be critical when the technology is applied to a new activity for which the institutional framework is not yet fully developed, i.e., when the framework is to be developed in parallel with the diffusion of the new technology. Therefore, the coevolution of social institution and technology can be expected when new technology is applied to newly created societal activities.

Three Dichotomies

Using the taxonomy of societal diffusion patterns shown in Table 6-12, we can develop *three* dichotomies. The first is between the installation of the computer and its utilization. Installation behavior follows conventional wisdom: the psychological approach is valid, the diffusion phenomenon is epidemic, and its time path follows the logistic curve. Therefore, the information-spreading mechanism determines the rate of computer installation.

When it comes to utilization behavior, however, the probit approach is valid, and Freeman's argument of techno-economic paradigm change holds: institutional inertia, rather than technological sophistication, is a determinant of diffusion. The second dichotomy, then, is between organizational and technical complexity. Organizational inertia is a determinant of the diffusion when applied to those activities that are organizationally complex. Organizational inertia is not a determinant of diffusion for those activities that are less complex organizationally but require technical sophistication. This is evidence that institutional, rather than technical, evolution determines diffusion patterns.

The third dichotomy is between old and new activities. Institutional inertia is not critical to diffusion when the technology is applied to a new activity. In this case, institutional coevolution will proceed in parallel with the diffusion of the technology. This third dichotomy establishes a new dimension in the study of technology diffusion in that it is not mentioned in Freeman's taxonomy, which is based on the hypothesis that institutional inertia deters the diffusion of technology.

ENVIRONMENTAL AND POLITICAL IMPLICATIONS

To date, we have discovered that regulation does not always discourage societal diffusion of new technologies but sometimes actually triggers the diffusion process. We have also found that one of the key determinants of technological diffusion is institutional inertia, but institutional inertia is lessened when the technology is implemented in parallel with the development of new institutional frameworks.

With these findings, we can broaden our discussion of the *coevolution* of technology and institutions to the global scene. The mechanism in which a new regulation accelerates the diffusion of technology may work for producing a new technology, even in a conventional industry, when a drastic change in regulation becomes socially indispensable. To consider this possibility, we will look at how the Japanese auto industry realized "technological profiting," when it developed new technology first to comply with the change in emission standards and then to accommodate economic concerns about fuel economy.

Political as well as economic systems can influence the diffusion of new technologies throughout a society. The collapse of communism and the end of the cold war are making it clear that technology drives the world political order. In this context, we will try to understand the impact of communism on technology diffusion. Under communism, what types of high technology were supported and what types were hindered?

Technological Profiting

As a technology diffuses throughout society, its widespread use may surpass a certain threshold of environmental capacity and thus add a new dimension of societal requirements for the technology to survive. Widespread use may lead to a situation in which the government has to enact new regulations, thus changing the rules of competition. This change may give newcomers an opportunity to become dominant technological players if they are aggressive enough technologically.[33]

If we look at the history of automobiles, we can trace the development of product, manufacturing, commodity, and societal technologies. The basic components of product technology were developed and perfected in the nineteenth century by such Europeans as Daimler, Benz, and Dunlop. This product technology was transferred to the United States and improved upon by Ford. Ford's major achievement, however, was, as everyone knows, the development of manufacturing technology, i.e., the conveyer belt system with interchangeability of parts. In the 1920s, consumers began to demand fashion and comfort in their cars. With these new market trends, Alfred Sloan of GM created commodity technology, including the closed model, full-line policy, consumer credit system, and a distribution system for used cars.

In the 1960s, when the diffusion rate of cars surpassed one car per two persons in the United States, people began to recognize a societal disadvantage of car usage—pollution. Thus, the development of a new technology was required to overcome this problem. This technology was quite different from the commodity technology developed by GM. We call it "societal technology" because it made survival of the car in the society possible. The Japanese contributions to developing this societal technology are obvious.

Two oil crises made society aware of another disadvantage to the automobile—the waste of precious nonrenewable resources. The increase in Japanese car exports in 1975 and 1979 coincided with the two oil crises and the increase in crude oil prices. Figure 6-5 makes visible the technological dynamics that occurred between regulatory changes in emission standards and the improvement in fuel economy.

As shown in the figure, the Japanese government successively raised emission standards for passenger cars, so that the allowable upper limits for emissions were drastically reduced between 1960 and 1978. To comply with emission standards, Japanese car manufacturers made substantial R&D investments, which began to produce improve-

FIGURE 6-5

Dynamic relation between emission regulation and fuel economy improvement

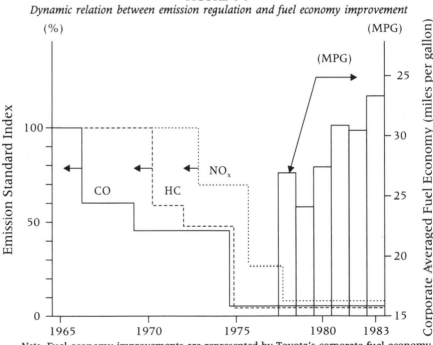

Note: Fuel economy improvements are represented by Toyota's corporate fuel economy.

ments in fuel economy after 1979. In fact, Japanese manufacturers were to discover that emission standards and improvement in fuel economy were two sides of the same coin as far as R&D efforts were concerned. In a public hearing before the Japanese Diet, Honda said that it was feasible to comply with the emission standards proposed by the government. This statement led to the government's decision to set the most severe standards in the world at that time, and the Japanese auto industry invested a vast amount of money and power in antipollution technology.

The requirement for emissions control turned out to be technologically equivalent to fuel efficiency. In order to comply with emission standards, Souichiro Honda, a founder of Honda Motor Co. Ltd., directed the company's researchers to focus on a deeper understanding of the burning process of fuel inside the cylinder.[34] When commodity technology was critical to competitiveness, this research area had been neglected by the auto industry worldwide because it did not contribute to a company's competitive edge. Honda's research strategy resulted in the invention of its *CVCC* engine. In hindsight, it is easy to see that Honda's accumulated knowledge about the burning process in the cylinder helped to improve fuel economy. Burning fuel cleanly is technologically equivalent to burning it "efficiently," especially in terms of designing the engine configuration.

The development of a new engine configuration is also related to the competition among Japanese car manufacturers at that time. Most auto manufacturers were newcomers to the industry. Honda had been manufacturing motorcycles, Toyota, textile machinery, Matsuda, tricycles for light-freight transportation, and Subaru and Mitsubishi had moved from their prewar business in airplanes to passenger cars. To a motorcycle manufacturer such as Honda, developing a totally new model was regular practice. Whenever one of its motorcycles does not win a competition, Honda will redesign the engine. This approach, however, was not the approach of firms that had always been in the auto industry. They would have preferred to develop a chemical catalyst that would remove the pollutant after the fuel was burned in the cylinder. Both Toyota and Nissan took this approach, as did U.S. auto manufacturers.

Regulatory changes, then, sometimes work to revitalize a conventional industry. New entrants from different industries can enrich technical variations created to comply with new regulations. By doing so, newcomers not only benefit from changes in regulation but also profit technologically. As a result of various societal technologies, including the automotive technology just described, Japan as a nation has come to consume less energy and produce fewer pollutants than the United States and Germany, as shown in Table 6-14. A study of the table reveals that only the emission of CO_2 is approximately the same for all three countries.

Political Implications

The diffusion rate of continuous casting methods in the Japanese steel industry has now passed 90 percent, but the technology itself was first developed in the Soviet Union. Why did the Soviet political system fail to facilitate the spread of continuous casting techniques, which are near the ultimate in technological development? In a communist system, central planning is absolute, and nobody can intentionally disrupt the system. Furthermore, production is not planned on a yearly basis, but almost to the day.

For factory managers, meeting these daily targets is their most important responsibility. To introduce continuous casting technology, factories must be closed down for a few years. As a result, factory managers responsible for maintaining daily production goals were opposed to change. In a nation whose planned economy regulates daily production volume at each factory, the more advanced a technology is, the

TABLE 6-14
International comparison in energy and antipollution efficiency, 1987

	Japan	U.S.	W. Germany
GNP per capita (US$)	15,760	18,530	14,400
Per Capita Commercial Energy Consumption (Eq. Kg Petroleum)	3,232	7,265	4,531
Total Primary Energy Requirement (TPER) (Mtoe)	372	1,865	272
SO_x Emission/TPER	0.35	11.49	9.75
NO_x Emission/TPER	3.57	10.71	10.97
CO_2 Emission/TPER	0.73	0.74	0.76

Source: Data on GNP per capita, The World Bank, *World Development Report* (Washington, D.C.: The World Bank, 1989); data for TPER, Organization for Economic Cooperation and Development, *Energy Balance* (Paris: OECD, 1989); data for SO_x and NO_x emission, OECD, *Energy Balance;* data for CO_2 emission, Y. Moriguchi and S. Nishioka, "Structure and Trends of CO_2 Emission in Japan and International Comparison," *Environmental Research Quarterly* no. 77 (1990).

lower the incentives are for factory managers to introduce such techniques because of the time involved in installing and learning the technology.

A nationally concentrated production system may appear to be advantageous for heavy industries because demand can be put together, but to allow the spread of the major technology changes, there must be some flexibility. A decentralized production system permits flexibility. The scattered and gradual realization of a long-term strategy of rapid progress in production efficiency and quality is possible because production planning is small scale and scattered.

It might seem that the demand for improved data processing technology would be greater in a central planning organization such as *Gosplan* than in a capitalist, decentralized system, but how does one explain the paradox that the technology for computers was discovered and developed in the United States? A national production planning system's demand for information technology is huge because of its massive information management requirements. The need for computers cannot be explained only by the need for large-scale data management. The real purpose of such a tool is based on the need for speed in handling data.

The increasing speed of information management has made it possible to refine production targets and to continuously eliminate the need for the creation of targets on a large, national scale. Thus, if we view technology as an intermediary variable, what looked to be paradoxical at first glance, is not at all. For this reason we have asserted that technology drives the world political order far more than the world political order drives technology.

CONCLUSION

Through empirical and quantitative studies of the diffusion of new technologies, we obtained new findings and clarified some of the effects of institutional and organizational influences on societal diffusion of new technologies.

While it has often been argued that institutional factors act as a barrier to technology diffusion, this chapter has shown that sometimes exactly the opposite occurs, with institutional response accelerating diffusion.

Widespread diffusion of facsimile communication in Japan, for example, began only after the revision of the Public Telecommunication Law. In this instance, regulation impeded the full exploitation of the potential of facsimile communication; after the changes in regulation were realized, facsimile communication diffused very rapidly. The rapid worldwide diffusion of facsimiles, however, was not a result of regulation. Rather, it occurred because the facsimile turned out not only to replace its predecessors, such as telex communication, but to dominate over them in every aspect of its functions. In other words, the facsimile overrode its predecessors. This accounts for its rapid diffusion in countries that have different cultural traditions and communicate in different languages.

The case study on the diffusion of CAT scanners in Japan shows that the relevancy of technology to institutional change can serve as an accelerator rather than a barrier. If a technology is both novel and intrinsically useful, as were CAT scanners for doctors, the introduction of the new technology triggers a chain reaction of institutional change that creates a favorable environment for the rapid diffusion of the technology.

The analysis presented in this chapter also has shown that it is important to consider certain organizational dichotomies that can influence diffusion. Significant dichotomies were found between the sophistication and simplicity of the technology and of the organization and, at a second level, between new and existing activities in the organization.

The quantitative analysis of the diffusion of the computer in Japanese prefectural governments reveals a more involved process of techno-economic paradigm changes. First, a distinction must be made between the installation and the utilization of the computer. The installation behavior follows conventional wisdom: the psychological approach is valid; the diffusion phenomena is epidemic; and the time path follows the logistic curve, which is the diffusion pattern of the conventional technology. For the utilization behavior, however, the probit approach applies and the theory of techno-economic paradigm change holds: institutional inertia, rather than technological sophistication, is a determinant of diffusion.

However, we must qualify the simple argument posed by the techno-economic paradigm change. Our study yields two dichotomies. The first is between organizational complexity and technical complexity. Organizational inertia is in fact a determinant of the diffusion of computer utilization when it is applied to activities that are organizationally complex. However, organizational inertia is not a determinant of diffusion when the computer is used for activities that are less organizationally complex but that require technical sophistication.

The second dichotomy is between applying a new technology to old activities and to new activities. Institutional inertia is a determinant when the computer is applied to conventional activities for which the institutional framework is already well established. On the other hand, institutional inertia is not critical to the diffusion of the computer when it is applied to a new activity for which the institutional framework is not yet fully developed, such as when the institutional framework is to be developed in parallel with the diffusion of the new technology.

NOTES

1. C. Freeman, *Technology Policy and Economic Performance* (London: Pinter, 1987), 60–79.

2. Ibid., 65.

3. G. Dosi et al., *Technical Change and Economic Theory* (London: Pinter, 1988).

4. F. Kodama, "Demand Articulation: Targeted Technology Development," in *Technik-Politik: Angeschichts der Umwelt-Katastrophe*, ed. H. Krupp (Heidelberg: Physica-Verlag, 1990), 273–94.

5. Ministry of Home Affairs, *Annual Report on Computers in Local Governments* (in Japanese) (Tokyo: 1985).

6. P. David, "Technology Diffusion, Public Policy, and Industrial Competitiveness," in *The Positive Sum Strategy: Harnessing Technology for Economic Growth*, ed. R. Landau and N. Rosenberg (Washington, D.C.: National Academy Press, 1986), 373–91.

7. F. Kodama, "Can Empirical Quantitative Study Identify Changes in Techno-Economic Paradigm?" *Science, Technology, and Industry Review* no. 7 (1990): 101–129. See also F. Kodama, "Paradigm Shift in Technology Diffusion: From Technical Change to Institutional Inertia" (paper presented at Stockholm Conference on Technology Management and International Business, Stockholm, Sweden, 1990).

8. F. Kodama, "Demand Articulation."

9. Ibid.

10. Michael Martin, *Managing Technological Innovation and Entrepreneurship* (Reston, Va: Reston Publishing Company, 1984).

11. M. Trajtenberg, *Economic Analysis of Product Innovation: The Case of CT Scanners* (Cambridge: Harvard University Press, 1990), 54–58.

12. D. Okimoto and A. Yoshikawa, *Japan's Health Care System: Efficiency and Effectiveness in Universal Care* (New York: Faulkner & Gray, Inc., 1993), 207–209.

13. Ibid.

14. E. Mansfield, *Industrial Research and Technological Innovation* (New York: W. W. Norton, 1968).

15. P. David, "Technology Diffusion, Public Policy, and Industrial Competitiveness."

16. E. Mansfield, "Microeconomics of Technological Innovation," in *The Positive Sum Strategy*, ed. R. Landau and N. Rosenberg (Washington, D.C.: National Academy Press, 1986), 307–25.

17. E. Rogers, *Diffusion of Innovations* (New York: The Free Press, 1962).

18. L. Nabseth and G. F. Ray, *The Diffusion of New Industrial Processes: An International Study* (New York: Cambridge University Press, 1974); Rogers, *Diffusion of Innovations;* V. Peterka, "Macro-Dynamics of Technological Change: Market Penetration by New Technologies," PR-77-22 (Laxenburg, Austria, International Institute for Applied Systems Analysis, 1977); J. C. Fisher and R. H. Pry, "A Simple Substitution Model of Technological Change," *Technological Forecasting and Social Change 3* (1971): 75–88; A. W. Blackman, "The Market Dynamics of Technological Substitution," *Technological Forecasting and Social Change* 6 (1974): 41–63.

19. P. Stoneman, *Technological Diffusion and the Computer Revolution: The U.K. Experience* (Cambridge: Cambridge University Press, 1976); P. Stoneman, *The Economic Analysis of Technological Change* (Oxford: Oxford University Press, 1983).

20. P. David, "Technology Diffusion, Public Policy, and Industrial Competitiveness."

21. A. Brian, "On Competing Technologies and Historical Small Events: The Dynamics of Choice Under Increasing Returns" (paper presented at the Technological Innovation Program Workshop, Stanford University, Stanford, Calif., 1983).

22. J. S. Metcalfe, "The Diffusion of Innovation: An Interpretative Survey," in *Technical Change and Economic Theory*, ed. G. Dosi et al. (New York: St.Martin's Press), 561.

23. P. David, "Technology Diffusion, Public Policy, and Industrial Competitiveness," 379.

24. P. David, "A Contribution to the Theory of Diffusion," Memorandum No. 71 (Stanford, Calif.: Stanford University, 1969); S. Davies, *The Diffusion of Process*

Innovations (Cambridge: Cambridge University Press, 1979); P. Stoneman, *The Economic Analysis of Technological Change.*

25. P. Stoneman, *The Economic Analysis of Technological Change.*

26. P. David, "A Contribution to the Theory of Diffusion."

27. T. Muraoka, "Analysis on Diffusion of Innovation" (in Japanese) (master's thesis, Saitama University, 1986); Kodama, "Can Empirical Quantitative Study Identify Changes in Techno-Economic Paradigm?"

28. P. Davis, *The Diffusion of Process Innovations.*

29. Ibid.

30. Ibid.

31. Ibid.

32. Although this cumulative normal time path is not dissimilar to the logistic curve, the logistic curve has longer tails.

33. F. Kodama, "Newcomers in the World Auto Industry" (paper presented to the International Auto Industry Forum, Hakone, Japan, 1984).

34. This story is based on the author's private conversation with Mr. Honda at Saitama University's Graduate School of Policy Science, 1985.

Epilogue
Techno-Globalism by Option Sharing

The emerging techno-paradigm shift does not always favor the existing Japanese system, especially in the case of large national science projects. Here, we are talking about those science and engineering projects that cannot be managed by scientists or corporations alone. These projects need direct government involvement.

For projects involving high-energy physics, space research, research into fusion energy, oceanography, and Antarctic research, even a single national government's involvement is not enough. International cooperation becomes necessary because none of us can easily afford the total cost, and all of us, large and small, share a desire for access to the best facilities for scientific research.[1]

In these areas, governments can play an important role in the creation of new science and technology, but it is a different role from that of corporations.

CONVENTIONAL STYLE

The cumulative nature of technological advance has been described by Nelson and Winter as following a *natural trajectory:* today's research produces successful new technology and the natural starting point for tomorrow's searches.[2] They discuss a "neighborhood" concept of a quite natural variety: once a system proves to be a success, it is necessary to make only minor changes. However, a set of technological possibilities sometimes consists of a number of different classes of technology. Within any of these classes, technological advance may follow a particular trajectory. At any given time, all R&D may be focused on one class of technologies with no attention paid to other classes.[3] These path dependencies, which are often involved in technology development, indicate the possibility that the system will lock into paths that are not globally optimal.[4]

The Manhattan Project is the conventional model for government involvement in the creation of new technology. According to Nelson, this model involves a willingness to make large early bets on particular technological options and force these through at very high cost.[5] International cooperation has followed this model and has been dominated by notions of cost sharing and task sharing. In the Foreword to this book, for example, Branscomb mentions cost sharing as practiced in the European Center for Nuclear Research (CERN), and task sharing as represented in the U.S. space station. However, dividing up costs and tasks suggests that an option has already been selected.

Certainly, this conventional model of government involvement is observable in Japan. Japan's total R&D budget for nuclear development is now far larger than any other country's, and comparing how this sum is being spent in Japan with how money is spent for nuclear development in other countries makes it clear that Japan has made a strong commitment to a specific technical option, the fast *breeder* reactor (FBR). Japan's FBR budget has become the largest in the world. Having reached that level, the market condition has become volatile, as have the domestic and international politics. Various countries, including the United States, the United Kingdom, Germany, and even France, have decided to curtail FBR programs.

However, all of these countries obviously have vested interests in keeping the FBR option open in case it proves necessary. Therefore, to break the prevailing stalemate in launching large science and engineering projects, as well as to realize an ideal of techno-globalism in the midst of the growing sentiments of techno-nationalism, we need to consider a new scheme of international cooperation.

JAPANESE NUCLEAR PROGRAMS

Japan's interests in nuclear technologies began in 1952 when the U.S. occupation ended and a ban on nuclear research was lifted. In 1955, Japan's Basic Atomic Energy Law was enacted, and the Japan Atomic Energy Commission (JAEC) was created in 1956 as the authoritative entity for policy decisions. JAEC's main task was to create a long-term development plan to promote domestic nuclear technological capabilities.[6]

JAEC published its plan for power reactor development in 1966. The plan, which is still the essence of Japan's existing nuclear reactor development programs, defined the fast breeder reactor (FBR) as the primary goal of advanced nuclear reactor development.[7] It is not surprising that a resource-poor country like Japan adopted FBR as its major goal, as in theory FBRs can produce more fuel than they consume. Furthermore, at the time of adoption, the world uranium resources were believed to be limited, and most other advanced nuclear nations took the same strategy.[8] The plan also called for the establishment of a domestic fuel cycle, which would include the reprocessing of spent fuel, which is essential to extract plutonium for FBR.

JAEC's plan was the first strong commitment to the development of "indigenous" technology in Japan. It was a break in the pattern of technology importation that had prevailed in Japan since the late 1880s. To achieve its goals, the government established the Power Reactor and Nuclear Fuel Development Corporation (PNC) in 1967 to develop FBR and its associated fuel cycle technologies.

Catching-Up Strategy

Although the goal of Japan's nuclear program was to establish indigenous technology, Japan's nuclear reactor development started with importation of foreign technologies. Although the Japan Atomic Energy Research Institute (JAERI) was established in the same year as JAEC to conduct basic nuclear R&D, the decision to import two research reactors and one prototype power reactor, all of them American-designed light water reactors (LWRs), was made as early as 1954. Although JAERI did build one research reactor, a heavy water reactor (HWR), with domestic technology, no other types of reactors were designed and built domestically.

Commercialization of FBRs was expected by the 1980s, but JAEC recognized that it would take a long time for Japan to develop FBR technologies, because the country had built only a HWR and had little experience in FBR.[10] To catch up with other countries, JAEC devised the following strategy: the experimental 50-megawatt reactor, *JOYO*, would be built first, while the parallel development of a prototype 280-megawatt power reactor, *MONJU*, would be pursued. Since the major mission of the newly created PNC was to build and construct these two FBRs by the mid-1980s, there was little time for PNC to explore various technological options.

PNC concluded international cooperation agreements with United States, France, Great Britain, and Germany to speed up the development process. Within Japan, all major nuclear companies (Toshiba, Mitsubishi Heavy Industry, Hitachi, Sumitomo, Fuji) were invited to join the JOYO project. These companies shared the responsibility for most of the design work and manufacturing of components.

Goals Achieved

The JOYO project was completed in 1977, finished its initial mission in 1983, and was converted to a larger fast-flux irradiation facility.

TABLE E-1
Total nuclear and FBR budgets in major countries

Country		Total ($billion)	FBR ($million)
Japan	(1991)	3.2	520
U.S.	(1992)	1.1	50
France	(1990)	2.0	61
Germany	(1991)	1.0	40
U.K.	(1990)	0.8	176

Source: Science and Technology Agency, *Nuclear Power Pocket Book* (in Japanese) (Tokyo: Japan Atomic Industrial Forum, 1992).

Notes: U.S. budget is the Department of Energy's nuclear R&D plus radioactive waste management budget. The FBR budget is "advanced reactor" budget. French budget is total civilian CEA budget. German budget is BMFT's total budget. UK's is UKAEA's total civilian budget.

The MONJU project was begun in 1967. Because of the lack of experience with fast reactors, the design work required many revisions during the 1970s. In addition, the different responsibilities assigned to participating companies made the technology transfer from JOYO to MONJU more difficult.[11] Despite the technical difficulties and delays, MONJU was completed in 1992, and it reached criticality in the spring of 1994 (a year later than it was planned).

Over the years, Japan's nuclear budget has increased drastically. Before 1967, the budget increase was moderate (¥9 billion in 1960 and ¥12 billion in 1965), but it jumped to ¥40 billion in 1970 and then grew sixfold in ten years (¥247 billion in 1980). In 1992, it reached ¥410 billion, making it one of the largest nuclear budgets in the world (see Table E-1).

Uncertain Future

PNC's original mission is now complete. MONJU will be the newest operating FBR in the world. PNC's FBR budget peaked in FY 1989 at ¥87 billion, and stabilized around ¥65 billion when MONJU construction was completed. The next step for FBR development is to build a demonstration power reactor (DFBR), but for several reasons the future of DFBR and the mission of PNC have become uncertain.

First, the commercial need for FBRs has become much more uncertain. Uranium resources are now believed to be plentiful, and the price of uranium is now the lowest it has been since the 1970s. On the other hand, construction costs as well as fuel cycle costs for FBRs are still very high.[12]

Second, existing FBR design and technology are no longer accepted as the best technical option for future nuclear development. For example, although a large-scale oxide-fuel FBR was believed to be suitable for commercial design, a small metallic-fuel fast reactor is emerging as a possible candidate. This type of reactor, called a liquid metal reactor (LMR), does not necessarily "breed" the fuel. Its design puts more emphasis on passive safety features rather than breeding capability. As waste management has become a critical issue to nuclear development, another fast reactor design that can "burn" or "transform" long-life radionuclides has become attractive. There is no clear consensus among nuclear experts today on the type of advanced reactor design that should be developed further.

Third, FBR uses plutonium, a material for nuclear weapons, as a fuel. Using plutonium as a fuel has increased political concerns about the risks of nuclear proliferation and possible diversion of fuel to nuclear explosives. Because of recent progress in nuclear disarmament, large quantities of plutonium may become available from dismantled nuclear warheads. This may increase political pressure to minimize civilian plutonium production. This political uncertainty and environmental concerns, which claim that plutonium is one of the most toxic materials in the world, have made the future even more uncertain for FBR programs.

Fourth, most advanced nuclear nations have either canceled or delayed FBR development programs. The United States canceled its demonstration project, the Clinch River Breeder Reactor, in the early 1980s. Germany's demonstration reactor (SNR-2) and prototype reactor (SNR-300) were halted in the 1980s. Great Britain has decided not to finance its FBR programs. Even France, which has been most aggressive in developing FBR, recently announced that it will not reopen its demonstration FBR (Superphenix) because of its technical problems and that there are no plans to build another.

Finally, the Japanese public attitude toward nuclear power has added further uncertainty to nuclear planning. According to a 1990 government poll, 90 percent of the public now feels uneasy about nuclear power, and 46 percent thinks that nuclear power in Japan is not safe. While a majority of the public (65 percent) still believes nuclear power is necessary for the country, the portion of the public that favors a larger role for nuclear power has dropped below half (48.5 percent) for the first time. Accidents in 1989 and 1991 have had a significant impact on public concern over nuclear safety. T. Suzuki has summarized the Japanese nuclear dilemma:

> To some Japanese leaders, any policy change seems like a threat to existing programs. But while consistency is certainly an element of success, other

considerations need to take priority. Japan's nuclear policy will remain credible only if it reflects new realities and reduces economic and political risks.[13]

A NEW COOPERATION SCHEME: OPTION SHARING

Nelson has noted that a government's aggressive support of engineering programs often involves "early commitment of governmental funds to a particular design," and he has pointed out that a government has a tendency to stick with a game plan despite growing negative evidence.[14] Indeed, every government is sticky about initiating work on new concepts.

Following this line of thought, Nelson goes on to assert "In the case of the *supersonic* transport, it is highly unlikely that Boeing would have persisted so long in pushing its swing wing SST design had the bulk of the funds been its own, and had it the ability to make the decision on its own."[15]

To confirm Nelson's concept of natural trajectory, we examined how industry has responded to government-led projects.[16] Japanese R&D statistics can supply industry's intramural expenditures by selected R&D objectives, nuclear development, environmental protection, and so forth, for example. These statistics disaggregate industry's expenditures into twenty-six industrial sectors.

By equating a technology with each industrial sector, we can measure whether or not there has been a shift in trajectory. In environmental protection at the local scale, for example, the R&D efforts are converging into specific sectors, while in nuclear development R&D efforts demonstrate a *cyclic* behavior in which diversion and conversion alternate. In other words, the environmental protection program is following a natural trajectory, while the nuclear program is experiencing several shifts in trajectories.

In order to accommodate the intrinsic dynamics of national programs, we are proposing international cooperation based on *option sharing*. Option sharing entails dividing up the burdens and responsibilities for pursuing each possible scientific and technological option in a given area. Recent news about cold fusion and warm-temperature superconductivity is a stunning reminder that science and technology entail a vast array of options and alternatives. A thorough search of all possible options, therefore, should be the main objective of future international cooperation. Conventional schemes of international cooperation, such as cost-sharing and task-sharing, have been developed by economists, not derived from the logic of science and technology itself.

Under option sharing, in the early phase of the development of large projects involving international cooperation, scientists in each nation would pursue the approach of their own choosing, which would be explored on an affordable scale. By international agreement, all information about each approach would be open to scientists pursuing complementary projects in other countries, and as each project matured, scientists could elect to work on the project of their own choice, regardless of national location.

Of course, this cooperation scheme should not permit one country to force the option it has selected on other countries. Each country should have the right to choose which option it wishes to pursue. Given the need to ensure that all possible options are covered, of course, there would have to be a certain amount of compromise and adjustment. In the case of projects like the super collider in which scientific value outweighs the merits of diversity, prior agreement would have to be sought for sharing costs and tasks to implement the scientific principles as a truly international facility.[17] However, the soaring costs involved in large engineering projects are due, at least in part, to the increasing number of options and to the pressure imposed on a single government to cover the costs of exploring all the options simultaneously. Only through international cooperation is it feasible to pursue all possible options.

Covering all possible options through international operation would have a profound effect on the development of technology. While science aims at absolute truth, technology aims at relative superiority. Determining the most meritorious technical option, therefore, is not possible unless all the options are demonstrated and compared. Option sharing should not be regarded as a country's relying on advances made by the competing projects of other countries. Instead, the other countries will provide a calibration of the state of the art of technical advance, with transparency provided through international cooperation.

Information sharing could be assured by allowing a free flow of researchers across national borders. After researchers had freely chosen the option they wished to pursue in accordance with their own views, convictions, and career objectives, they would work in the country pursuing that option. Once the best option had been determined, researchers would return to their respective countries, thus ensuring that information on the option will be disseminated throughout participating countries.

Through option sharing, it is possible to resolve the inherent tension that exists between international cooperation and national autonomy. In Branscomb's phrase, through the principle of *cooperate and compete*, nations in the industrial world may capitalize on parallel interests. There are growing fears that the shift toward technological protectionism will turn into a minus-sum game for the world as a whole.[18] It can be said that only through option sharing can a plus-sum game be assured. In a world in which "techno-nationalism" is the prevailing mood, international cooperation through option sharing may offer the breakthrough that can make the ideal of "techno-globalism" the new reality.

NOTES

1. L. Branscomb, "The Road from Resentment to Understanding in U.S.-Japan Science and Technology Relations" (paper presented at the 100th Anniversary of Electrotechnical Laboratory, Tsukuba, Japan, 1991), E-27.

2. R. Nelson and S. Winter, *An Evolutionary Theory of Economic Change* (Cambridge: Harvard University Press, Belknap Press, 1982), 257.

3. Ibid., 262.

4. R. Cowan, "Nuclear Power Reactors: A Study in Technological Lock-in," *The Journal of Economic History* 50, no. 3 (1990): 541–67.

5. R. Nelson, *The Moon and the Ghetto* (New York: Norton & Company, 1977), 119.

6. For a more detailed history, see T. Suzuki, "Japan's Nuclear Dilemma," *Technology Review* October (1991): 44–49. This section is based on Suzuki's paper "Need for Strategy Shift in Japan's National Big Project," which was prepared in August 1992 for the Japanese Commission on Industrial Performance, organized by the Japan Association of Techno-Economic Society (JATES).

7. *Basic Course of Reactor Development* (in Japanese) (Tokyo: Japan Atomic Energy Commission, 1966).

8. In fact, JAEC dispatched an ad-hoc studying commission to Europe and the United States to learn about other countries' strategies before reaching this decision.

9. Although the first commercial power reactor was the gas cooled reactor (GCR) imported from Great Britain, LWR became the dominant reactor in Japan. The technologies were licensed from two U.S. reactor vendors, General Electric and

Westinghouse. The decision to import LWRs from the United States was heavily influenced by the U.S. policy decision to supply enriched uranium, which is the primary fuel for LWRs. Japan was then trying to develop a heavy water reactor, which does not require enriched uranium.

10. JAEC also announced plans to develop an advanced thermal reactor (ATR), which was based on Japanese HWR technologies. The ATR was viewed as an interim reactor. JAEC did not expect to build an experimental reactor for the ATR and thus started with the prototype reactor, FUGEN. Industry resisted this ATR development since it could diffuse R&D efforts and slow down FBR development. Industry even suggested importing FBR technologies to speed up FBR development.

11. The prime contractor for JOYO was Toshiba, which was responsible for the reactor design. To share the technology widely, the reactor design work for MONJU was first split among four vendors and later shifted to Mitsubishi Heavy Industry, which was the prime contractor for MONJU.

12. For example, MONJU's unit construction cost (¥ 2,400,000/kw) is more than seven times that of commercial LWR unit construction cost (¥ 300,000–350,000/ kw). Utilities are insisting that the next DFBR should not cost more than twice the cost of LWR.

13. T. Suzuki, "Japan's Nuclear Dilemma," 49.

14. R. Nelson, *The Moon and the Ghetto,* 119.

15. Ibid., 120.

16. F. Kodama, "Technological Entropy Dynamics: Towards a Taxonomy of National R&D Efforts," in *Measuring the Dynamics of Technological Change,* ed. J. Sigurdson (London: Pinter Publishers, 1990), 146–67.

17. Branscomb, "The Road from Resentment to Understanding in U.S.-Japan Science and Technology Relations," E-27.

18. R. Reich, "Beyond Free Trade," *Foreign Affairs* 61, no. 4 (1983): 773–804.

Index

About the Author

Fumio Kodama is a professor of science, technology, and industry at the Research Center for Advanced Science and Technology, The University of Tokyo. Previously he taught at the Tokyo Institute of Technology and at Saitama University. He also served as a visiting professor at Harvard University, Stanford University, and Hamilton College.

In addition to teaching, Kodama has worked as director-in-research of the National Institute of Science and Technology Policy, as an administrator at MITI, and as a research fellow at the Institute of Systems Research, Heidelberg. He serves on the board of directors of the Engineering Academy of Japan and on several advisory committees to the Japanese government, including committees at MITI, the Science and Technology Agency, and the Economic Planning Agency.

The author of many publications on science and technology policy, Kodama received the 1991 Sakuzo Yoshino Prize for *Haiteku-no-Paradaimu (The Paradigm of High Technology)* as well as the 1991 Science and Technology Minister's Award for Research Excellence. Kodama is co-author, with Lewis M. Branscomb, of "Japanese Innovation Strategy," published in the Harvard University Center for Science and International Affairs Occasional Paper Series.